KU-633-602

BETH CHATTO

A life with plants

BETH CHATTO
A life with plants

CATHERINE HORWOOD

PIMPERNEL
PRESS LTD
www.pimpernelpress.com

Pimpernel Press Limited
www.pimpernelpress.com

BETH CHATTO
A Life with Plants

Design by Anne Wilson
Typeset in Lunaquete

A catalogue record for this book is available from the British Library.

ISBN 978-1-910258-82-8

Printed and bound in China
By C&C Offset Printing Company Limited

9 8 7 6 5 4 3 2 1

Author's note: for brevity when writing her diaries and notebooks, Beth frequently shortened names to a letter such as 'A' for her husband, Andrew. Similarly place and plant names were often abbreviated. In editing them, I have used the full names of both people and plants.

Beth called Christopher Lloyd either Christo or 'C'. I have used Christo throughout as that is how he was known universally by his close friends – and also by many others.

Where names are marked with an asterisk, brief biographical notes can be found at the back of the book.

In a few cases, the botanical names of plants have changed since her diaries and notebooks were written. In these instances, I have retained Beth's originals. The index includes current names.

Contents

THE BETH CHATTO GARDENS
THE BETH CHATTO GARDENS

Foreword by Julia Boulton

MY GRANDMOTHER, Beth Chatto, was a remarkable character. You could not fail to admire her extraordinary vision and energy in creating a garden from a wilderness. Ahead of her time in propounding the ecological concept and practical application of 'right plant, right place', she was also a brilliant communicator both in person and in the written word.

Her professional career took off when she was in her forties, resulting in: exhibiting at RHS Chelsea (winning ten Gold Medals), writing eight books and numerous columns, international lecturing, the VMH and an OBE . . . She inspired me and the rest of her family, as well as her staff – and gardeners and plant lovers all over the world – to plant according to the rules of nature.

Her story is inspirational and I'm so glad that she took the time to share it with Dr Catherine Horwood, who, as well as spending many hours with her, was given access to Beth's archives, including comprehensive diaries and meticulously kept correspondence.

Beth left a large footprint: her books, her gardening philosophy, her beautiful gardens and extensive plant nursery, and most recently her Education Trust. She cannot be replaced but her family, the loyal team she built, her huge circle of horticultural friends from far and wide and I proudly continue her traditions and her work through the business she created and the charity she formed.

I hope the biography gives you an insight into my dear grandma's drives and motivations. Also a huge thank you to Dr Catherine Horwood, who has worked so diligently on this book and has generously gifted half of the royalties to Beth's education charity to help provide horticultural inspiration, enjoyment and education to people of all ages.

Introduction

In 1989, at the end of an international symposium in Melbourne, Australia, the dozen or so speakers were called to the stage to receive a thank-you gift. When the organiser came to the last name, he simply said 'Beth'. There is only one woman in the horticultural world known instantly by her first name. There is only one Beth.

Beth Chatto's gardens and nursery have influenced gardeners across the world. Her writing continues to guide new generations. But this book is not about Beth's gardens nor about the plants she popularised. It is about Beth as a person – though, of course, in many ways the three are inseparable.

In 2011, Christopher Woodward, director of the Garden Museum, in Lambeth, London, asked me to help Beth prepare her archive on the development of the Beth Chatto Gardens and nursery for its eventual transfer to the museum. Without hesitation, I agreed. I had been a fan of Beth's since the 1970s when I was among the eager crowds clustered round her stands at the Chelsea Flower Show, and I had written about her in my book, *Gardening Women* (2010).

I would arrive at Elmstead Market once or twice a month to be presented by Beth with an envelope of new finds, with instructions to throw them away if they weren't useful. They always were – a letter from Christopher Lloyd one week, a plant list the next. And then there was the time spent with Beth – chats to start with, longer talks later.

It soon became obvious that, quite apart from the history of the gardens and nursery, Beth's life was also a story that had to be told. Encouraged first by Tricia Brett, Beth's PA at that time, and later by her granddaughter Julia Boulton, now managing director of the Beth Chatto Gardens, in 2013, Beth asked me to write her biography. Once again, I leapt at the opportunity.

I have used three main sources. The first is the vast archive, both personal and professional, Beth had accumulated over half a century. It comprised thousands of items: press cuttings, articles written by and about Beth, photographs, slides,

Beth Chatto's professional archive was one of the first to be donated to the newly extended Garden Museum at Lambeth in 2017.

correspondence, stock books, seed lists, and her handwritten book manuscripts (she never learned to type) together with the subsequent typescripts, heavily marked with her edits and revisions.

Most important were her private diaries, a near complete run from 1963 into the 1990s, and her travel notebooks. Throughout her working life, although with some gaps, Beth kept a daily diary and detailed notebooks on her later trips abroad. Her diaries range from tiny pocket books in the 1960s to large foolscap size by the 1990s. But they always had one page per day with her notes on the weather scribbled at the top, often her supper menu at the bottom, and in between, all the day's events, from plants she had propagated to when – and by whom – she had had her hair permed. Her handwriting was always elegant, although the legibility varied with her mood. Her travel notebooks were almost all in cheap Silvine notebooks, with bright red covers that would be instantly recognisable to a generation who bought them for pennies from Woolworths. I was given access to all these often extremely frank documents by Beth, who was fully aware how useful they would be to a biographer.

Like most diary writers, Beth noted the mundane events of the day but also sometimes her most intimate thoughts. Her husband, Andrew, read neither her diaries nor her travel notebooks, which Beth used as an aide-memoire of people met, places visited and, of course, plants seen. There are long breaks – months, the odd year missing – either lost or, more likely, from times when Beth was too distracted by the gardens, other writing, visitors, life, to keep them up. For, as she wrote in *Beth Chatto's Notebook* in 1988, 'even keeping up a scrappy diary becomes difficult when the sap rises.'

Secondly, I watched and listened to film, television and radio material relating to Beth from across the years. In 2002, she was interviewed about her life and work by Louise Brodie from the British Library as part of an oral history project on horticulture. The tapes, lasting around twelve hours, were a valuable source, especially early in the project.

Finally, I have interviewed dozens of people who knew Beth: her family, friends, horticultural colleagues and the large extended 'family' of staff and former students – those who worked for her over the years, several of whom still do. Most important of all, of course, was the time I spent with Beth herself over her last years as we got to know each other.

A biography is like an incomplete jigsaw puzzle. You fit together as much as possible but there will always be gaps that can't be filled. Beth's mind was sharp until the end. Inevitably, there were things she couldn't, or wouldn't, talk about because of memory or reluctance. What never left her was the steely determination to get things right. I hope she would feel that I have done that in telling her life story.

1
Under the gooseberry bush

An interesting wedding was performed by the vicar Rev. Julian J Butler on Saturday August 7 when Miss Betty Diana Little, only daughter of Mr and Mrs W G Little of the Police House, Elmstead, was married to Mr Andrew Edward Chatto . . . The bridegroom is a well-known Essex fruit farmer at White Barn Farm, Elmstead. The bride, who was given away by her father, was attired in a dusty pink dress with brown accessories, and carried a bouquet of red roses . . . The honeymoon will be spent touring Wales, the bride's travelling dress was a grey flannel suit with rust accessories.

THIS PRESS CUTTING from an unnamed local Essex newspaper in August 1943, faded from being stored for years in a scrapbook, the Order of Service from that 'interesting' wedding, and a few black and white photographs are all that remain from that day. It was, after all, a war wedding – nothing very unusual in that. It was the couple who made it 'interesting' – the 'well-known' and well-heeled Essex fruit farmer marrying the twenty-year-old daughter of a village policeman, fourteen years his junior, may have set tongues wagging in Elmstead Market, then a small village, six miles east of Colchester, Britain's oldest recorded town.

Their lives and backgrounds could not have been more different. Beth, or Betty (or Bessie) as she was still known, grew up in a house without running water or electricity in rural Essex. Andrew Chatto, the middle-class publisher's son, brought up in the wealthy Hertfordshire town of Radlett, had already been to California by the time he was a teenager. It was not an obvious match.

Mr and Mrs Andrew Chatto leave the church of St Anne and St Laurence, Elmstead Market, on 7 August 1943, with the former Betty Little clutching her bouquet of red roses.

Beth and her twin brother, Seley, were born on 27 June 1923, at their parents' home in Tye Green, a hamlet of the village of Good Easter, seven miles north-east of Chelmsford – or 'the sticks of Essex' as Beth called it. The pair were christened Betty Diana and William – after his father – Seley, an unusual name for which there seems to be no family connection or explanation but by which he was always known.

Their father, William Little, was a village police constable, not a very onerous job in the crime-free Essex countryside but it came with some status, together with a house. William would have been relieved to receive the posting to Good Easter from a larger station at Grays in Essex just six weeks before the twins were born. A keen churchgoer and gardener, he was a tidy, organised man who polished not only the brasses in the church but also the spades and forks in his garden shed. He was the youngest of seven children.

His father, also William, was illiterate and started working, as his father had done before him, as a farm labourer at a time when, in East Anglia in particular, most people worked on the land. The school-leaving age was legally twelve, but little was done to encourage education in such rural communities. However, William senior did break away from agricultural work and went on to run a horse-drawn bus business in Writtle, near Chelmsford, before the Great War. But by the 1920s, motor cars and buses had made horse-drawn buses redundant and William junior had to look elsewhere for work. Beth's father, having learned to drive, found employment as a chauffeur for a local doctor before enrolling with the police in 1921.

Beth's mother, Bessie, was a Warwickshire girl, born in 1888 in Wootton Wawen, just north of Stratford-upon-Avon. She was in service before the First World War as a maid-of-all-work for a vicar, his wife and young son in Tamworth, Staffordshire. This was one of the hardest domestic roles since no other help was kept. At the outbreak of war, she enlisted to train as a nurse and was posted to Essex. She never lost her Midlands accent and was a good talker in a soft, mild way. She soon became engaged to William's older brother, John Little, who was serving as a stoker on a submarine. He was killed when it was lost in a tragic collision off the Essex coast near Harwich on

LEFT ABOVE 'Chestnuts', Tye Green, Good Easter, Essex, the birthplace of Beth, then Betty Little, and her twin brother, Seley, on 27 June 1923.
LEFT BELOW William and Bessie Little with twins Betty (right) and Seley (left).

19 January 1917, but Bessie stayed close to the Little family after John's death. Before long, she started 'walking out' with John's younger brother, William (despite being twelve years older than him). In April 1922, his appointment as a police constable was confirmed and on 7 October 1922, they were married in Hammerwich church, near Stafford, with both families in attendance. Beth and Seley were born eight and a half months later when Bessie was thirty-five years old.

It was a happy marriage, with the Littles retaining their sense of humour and fun while always keeping to strict social standards in and out of their home. Life was simple, based on a love of the country with long walks and respect for the seasons. Beth rarely saw her maternal grandmother, Rhoda, in Warwickshire, but Bessie always claimed it was from Rhoda that she got her interest in wild plants and herbs.

When Beth – or Bet, as she was known in the family – and Seley were still babies, William passed his sergeant's exam at the second attempt. Despite this, he remained a constable throughout his career with the police, since promotion was often more to do with vacancies than ability. In January 1926, he received a new posting and the family moved to Great Chesterford, near Saffron Walden. Once settled, the Littles soon acquired a pedigree golden retriever called Gypsy. They bred from her and, for a short idyllic time, the house always seemed full of puppies.

When Bessie became pregnant again at the impressive age of forty-one, Beth and Seley were sent away to stay in Warwickshire with Rhoda. Their return, several months later, was to a sadder house. Their mother had once again given birth to twins, also a boy and a girl. But the little girl, named Mary, lived for less than twenty-four hours. And, not long after, Gypsy, the much-loved family dog, was hit and killed by a car.

There were to be no more children or dogs. Beth was left as the girl between two boys. She was a happy little girl but occasionally she was jealous of them, feeling that they were their mother's favourites. Bessie Little was a firm woman with traditional ideas. Throughout her childhood, Beth was expected to help with the household chores while Seley and David were never asked or made to do anything. Later in her life, Bessie suggested that perhaps her determined daughter should have been the boy, and the more delicate Seley, the girl.

Life at home was never dull. The family lived in an old chapel that had once been two cottages knocked into one. It was a rambling old place, and cold. The

The Little home at Great Chesterford, Essex, where Beth was given her first patch to garden.

Beth and Seley with Gypsy, the family dog, and some of her puppies, at Great Chesterford.

picturesque well-head in the garden was for real: the water had to be dragged up and brought into the house. The village had no mains water, gas or electricity, so everyone lived with oil lamps, candles and well water. The children were used to it – they knew no other way – and they enjoyed watching the frost pictures on the windows in winter. There were two coal stoves to keep them warm: a cooking range in the kitchen, and a Tortoise stove in the living room. These Tortoises, black, cylindrical cast-iron stoves, local Essex products, were widely used as a low-cost form of heating. Beth and her brothers would lie in front of their Tortoise, enjoying the warmth while reading their books.

Reading, play and gardening occupied their free time but there was also the routine of strict religious observance. William and Bessie were regular churchgoers supporting their local Anglican church in Great Chesterford. For the children,

The Littles were keen gardeners at Great Chesterford. Beth would sit on the flint wall watching the ducks in the fields beyond.

singing in the choir, Sunday school and church services were a big part of the weekly round.

Although it was the Depression era and wages were low, it was a happy, carefree time for the children. The games they played hadn't changed much since Victorian times. Beth was useless at trundling an iron hoop but loved playing with marbles, spinning a wooden top with a piece of string or driving her mother mad by playing ball on the side of the house. This was a rural world with few cars on the roads, no tractors on the fields, no herbicides or pesticides.

The Little children had freedom to roam, and they did, clambering through the hedgerows, exploring the woods and the water meadows, and fishing in the streams. The lane to the village ran just outside their garden wall. Beth would sit on the flint wall watching the ducks in the meadow opposite, twenty white ducks who came down from the farm every day to dabble in the shallow puddles. In the evening in late spring, the whole family would sit on the wall watching the owls feed their young in the elm trees opposite.

There were no cars in the village. Everything for the baker and grocer was brought in by cart, as was the coal. Nearly everyone worked either in or near the village, on the farms or in the dairies where Beth went to watch the milk being skimmed and the butter churned.

William and Bessie Little were keen gardeners and grew all their own vegetables. The garden at Great Chesterford, designed by William, was laid out in the traditional cottage style – an informal mix of fruit and vegetables, with flowers to fill in the gaps. It sloped down towards the meadow across the lane, surrounded by flint walls which were covered with rambling roses and honeysuckles. There were greengages, full of wasps when ripe. Beth used to climb on the wall and walk along it to reach them. The garden seemed full of forbidden fruit – gooseberries, raspberries, blackcurrants and redcurrants draped, as was the tradition, with old lace curtains to keep off the birds and, less successfully, children. Aged five or six, Beth lay beneath the gooseberry bushes looking for those that had been missed in earlier pickings, waiting for the thick, sharp skins to become transparent-thin, or plum red, to burst in her mouth with unforgettable flavour and sweetness.

Unlike her brothers, Beth took to gardening and loved it. She was always keen to help her father, learning from him how to string onions, lift potatoes and sow vegetable seeds. As a reward, he gave her a patch to garden, a damp, shady corner near the water butt, where she planted some lily of the valley. Soon she added snowdrops and ferns.

The primary school in Great Chesterford was a typical Church of England village school. On her first day, Bessie took Beth and Seley into what seemed like an enormous room. All the little desks were at one end; the inevitable Tortoise stoves were in the middle and the activity area was at the other end. The youngest children sat playing with plasticine and Beth and Seley were given some to fiddle with while the teacher talked to their mother. Beth was already a voracious reader and the plasticine did not amuse her for long. She was a bright child who enjoyed learning and was to love school. The teachers soon found that she had already read everything available to the first year and pushed her up into a higher year group.

Beth (Betty Little) in 1933, aged ten.

Most locals worked on the land so, even in the classroom, many subjects revolved around the seasons. The children were taught to grow peas and beans on blotting paper but Beth was more ambitious and wanted to grow annuals such as asters. These were not perhaps the best choice for her shady corner at home but she still found it magical to see the first shoot and root going up and down inside the jar.

Beth's love of school convinced her from an early age that she wanted to be a teacher when she grew up. Even at ten years old, sitting for her photograph, she looks out with determination in her eyes, arms crossed defiantly across her cardigan, dress buttoned up to a smart lace collar. A slight enigmatic smile on her lips, she was already hard-working and ambitious, wanting to experience life, to travel, to meet people. 'I had no idea where or how I would go but I somehow felt I was driven or had the energy to discover a wider world.'

IN MARCH 1935, when Beth was twelve, her father was given a new and final police posting to Elmstead Market, on the other side of Essex, six miles east of Colchester. For Beth, it felt like the end of a golden childhood to be leaving Great Chesterford, the home and garden she had loved, for a new-build police house by a busy road leading to the popular seaside resorts of Clacton and Frinton, and the port of Harwich. In later years, she hated to admit that she had ever lived in that house and would never talk about it even to her grandchildren.

But the move did bring Beth the opportunity to go to the excellent Colchester County High School for Girls. Ambitious and bookish, she loved the high standards and challenges that it set. Her talents did not go across the board: she hated maths and did not do well in it. But she got good marks in English, and adored poetry and other literature, and writing essays. She was always studying others' writing, feeling that it helped her with her own.

The school served a farming community as well as the town population but, despite her tomboyish tendencies, Beth was reserved and, a latecomer to the class, arriving halfway through the school year, made few friends. With the fear of war looming, the girls' lives became more restricted. The headmistress, Miss King, moved the younger classes into Grey Friars, an eighteenth-century house not far from the Castle in Colchester with tennis courts at the back. Below them were fruit trees and a vegetable garden which were used to feed the school. Against the wall at the bottom of the garden were around twenty or thirty small plots where girls who were interested could make a garden. Inevitably, Beth was one of them.

Although they each had to share their small plot with one other, the girls could then plant whatever they wanted. With a packet of seeds from Woolworths costing just a halfpenny or a penny, they would grow annuals such as poppies, clarkia and night-scented stock. Left to their own devices, the girls soon learned from each other what was a weed and what wasn't, pulling out the tiny weeds from between the flower seedlings. If one pupil had done better than everyone else, she was allowed to wear a special gardening badge for a year. This did not escape Beth's notice, and before long she was wearing one.

THE OUTBREAK OF WAR in September 1939, Beth's final year at Colchester High School for Girls, meant inevitable changes in the Little household. William Little's status as the local bobby gave him a leading role in the Home Guard, set up in 1940 as a defence against possible invasion. As a coastal county in south-east England, Essex was in the front line and Colchester was a garrison town and the base for several additional battalions, as well as its home division of the Fourth Infantry. Dozens of concrete pillboxes dotted the countryside, including around Elmstead Market, midway between Colchester and the coast.

The Littles' police house home became a meeting point for everyone involved in local defence. With an atmosphere reminiscent of the BBC's Home Guard programme, *Dad's Army*, the Elmstead villagers pulled together. Bessie supplied coffee and home-made cakes, often in the middle of the night, to sustain spirits among the helpers.

Among them was local fruit farmer, Andrew Chatto, then thirty-one years old. Conscription meant that there were few men of his age around. He was excused military service since anything to do with food production, such as fruit farming, was a reserved occupation. As well as managing the farm, Andrew became an air raid warden and was soon a familiar face at the police house. Andrew, along with William and the local schoolmaster, also organised temporary homes for children evacuated from London and billeted in the village for safety.

Young David Little, by then a bright twelve-year-old, was delighted to have the chance of an occasional game of chess with the fruit farmer to while away the long hours. His seventeen-year-old sister, meanwhile, sat at the other end of the kitchen table assiduously doing her homework. Beth's life was sheltered, not just by the effects of war on the village but by her parents' strict values. The children were not allowed out in the evening except to go to church or choir practice and rarely mixed with other families in the village.

The arrival of Andrew Chatto into the Little home was a serendipitous result of the social upheaval of war which allowed for a mixing of classes and sexes in a way that would have been rare before. Beth's background was very different from Andrew's privileged upbringing, but there were similarities: shared values and a love of natural things and the countryside.

ANDREW CHATTO, like his father before him, was named after his grandfather, co-founder of the famous publishing firm Chatto & Windus. The company was successful and the Chattos lived a comfortable life. Family photograph albums show a large, prosperous Victorian and later Edwardian family, dressed in the height of fashion and with sufficient funds and sophistication to travel widely across Europe and to the United States. Andrew's father, who had gone into the family business, had surprised his parents by marrying an American, Elisabeth Boote, from New Jersey. Elisabeth was the only girl among five brothers and already in her thirties when they married. Both were by nature rather shy and retiring, characteristics inherited by their only son.

Andrew, known as 'Dan' in his family, benefited from the success of his grandfather's publishing business and grew up in relative wealth in Radlett, Hertfordshire. However, Elisabeth Chatto felt lonely and isolated in England, her health was poor and she never felt comfortable with the British way of life. In 1921, her husband resigned from the family business and took her and their son, then twelve, to live in southern California where one of her brothers had a ranch growing oranges. They sailed first class on the Cunard steamship RMS *Carmania* to New York and then crossed the United States by train, via the Grand Canyon, to Laguna Beach in Orange County, where there stretched mile upon mile of orange groves.

When this slightly awkward English boy arrived in this exotic landscape, he rushed outside to explore the surrounding hills. He was thrilled by the real-life cowboys and Indians still much in evidence, later telling stories to his daughters of how tough they were. He took photographs of the surrounding countryside and bought postcards of famous natural landscapes such as the Grand Canyon, carefully sticking them into albums he kept throughout his life. He was particularly mesmerised by the plants he found growing wild all around. His first question was, 'How did ceanothus and orange poppies get here from Radlett?' The answer, that they were native plants in California, sowed a seed in his mind. The story of this memorable trip, much told in the Chatto family, became the major influence in Andrew's life. He continued to question where the flowers and shrubs growing in his suburban garden in England had come from; it became a lifelong quest.

The Chattos stayed in America for two years, with Andrew going to school in a wooden shack alongside the immigrant families pouring into the state. But eventually the Chattos decided that they ought to return to England to give him a rather less primitive education. After being sent to a traditional British private day school near his home in Hertfordshire, Andrew went to Wye College in Kent, an agricultural training centre, to study farming; despite his scholarly bent, he had no interest in going into the family publishing business.

In 1930, the family moved to Braiswick, a suburb of Colchester. Andrew's father had a house built for them, 'Weston', and also bought land for Andrew, now twenty-one, to turn into a fruit farm – not an orange grove like his uncle's farm in California, but an orchard of hardy English dessert apples and pears. The 100-acre White Barn Farm was at Elmstead Market, well away from the comfortable, residential area of Colchester where Andrew lived with his parents. It was soon planted with fruit trees. However, Andrew found that it was a harsh environment in which to produce good fruit. They had a wonderful flavour but Elmstead, with its sandy, gravelly soil, needed constant irrigation. Despite his deep love for and knowledge of nature, Andrew was not a natural farmer. He was not enthused by the day-to-day running of the business and much preferred his reading and scholarly research. By the time he met Beth, he already had an encyclopaedic knowledge of plants and botany.

In the summer of 1940, just after her seventeenth birthday, Beth passed her School Certificate in six subjects, with top marks in English Literature, History and Geography. Her results were good enough to win her a place at Hockerill Teacher

Training College in Bishop's Stortford, Hertfordshire. Established in the mid-nineteeth century as the 'Diocesan Institute for the Training of Schoolmistresses', the college was still imbued with the religious values that Mr and Mrs Little wanted their precocious daughter to retain.

The college was a former priory, centred around a quadrangle of attractive old buildings. But on 11 October 1940, soon after Beth arrived, part of the student accommodation was bombed – the only direct hit on Bishop's Stortford during the war. Three students were killed. The girls were billeted out in small groups of a dozen or so in old Victorian houses around the town with a couple of elderly female lecturers to keep an eye on them.

The next day, instead of starting lessons, the girls were somewhat shocked to be sent out to collect potatoes from the college's 2-acre plot. With labour short because of conscription, everyone was expected to muck in to gather food for the college canteen. A machine dug them up but the new students had to follow and pick them up in buckets. Although Beth was a country girl and an experienced gardener, she never forgot the back-breaking pain of that first day.

Nevertheless, the college garden was her great joy. Alongside it there was a little primary school attached to the college with around two dozen eight- to eleven-year-olds on whom the trainee teachers practised their skills. One day, Beth and two other students were offered these children for a month and allowed to do whatever project they liked with them. One of Beth's friends, Mary Martingale Taylor, was a musician who played the violin; the other, Margaret Versey, was literary-minded, while Beth's interest was naturally gardening. She took the lead and it was soon decided to make the garden the centre of the project, with all the lessons designed around it. For maths, they had the children mapping the garden, making right angles and recording all the measurements. Mary, the musician, organised dancing and songs related to the garden and they all encouraged the children to think poetically about the flowers and the plants. The experiment was a great success.

Beth with Margaret Versey (left) and a schoolfriend (centre) at Hillside House, the wartime accommodation of Hockerill Teacher Training College, Bishop's Stortford, July 1942.

With Beth away in Bishop's Stortford, she and the new family friend, Andrew Chatto, began a regular exchange of letters over everything that was happening at her college and his farm. One of the first questions she asked Andrew was about a flowering shrub she had spotted in the college garden, bare twigs covered with little, sweetly scented yellow flowers like wood shavings. Back came the answer: the Chinese witch hazel, *Hamamelis mollis*. She relished having such a fount of horticultural knowledge on hand.

At college, Beth spent much of her time in the library, reading not only her set texts but also early books on ecology and on how plants adapt themselves to their conditions. She was already aware of the differences between the flora on the chalk land of her childhood village in west Essex and that around her new home on the dry sandy soil of Elmstead Market.

As part of their teaching degree, all students were required to do a major project in one of their specialist subjects. Beth's were literature, biology and botany. Inevitably, she chose botany for her project. In the summer months, she and Andrew regularly swam in the tidal rivers of the Colne estuary. When there, Beth's eyes were drawn to the varying colour patterns of the salt marshes. According to the concentration of salts round the low-lying pools, there were different coloured bands of plants which had adapted to the places where salt had accumulated. These changed further upriver where the fresh water came in from the land behind. Knowing about Andrew's trip to California and his interest in ecology and mapping areas of plant associations, she plucked up courage to ask him if he would help her with her project, a study of the salt marshes. The answer, not surprisingly, was positive.

Armed with rulers and coloured pencils, they began the survey. It took them two years to complete. Andrew's instructions were clear: 'Select site(s) – geographical location, climate and exposure, local soils, draw map.' Starting from the saltiest areas and going back to where the fresh water was coming in, they marked out 6-foot-square quadrats. These in turn were divided into 1-foot-square quadrats with every plant in every square drawn and coloured. Dominant plants were represented by solid colour, less dominant ones by alternate levels of colour of plants represented. Once laid out on paper, the repetition and then the loss of some plants and the development of others became clear.

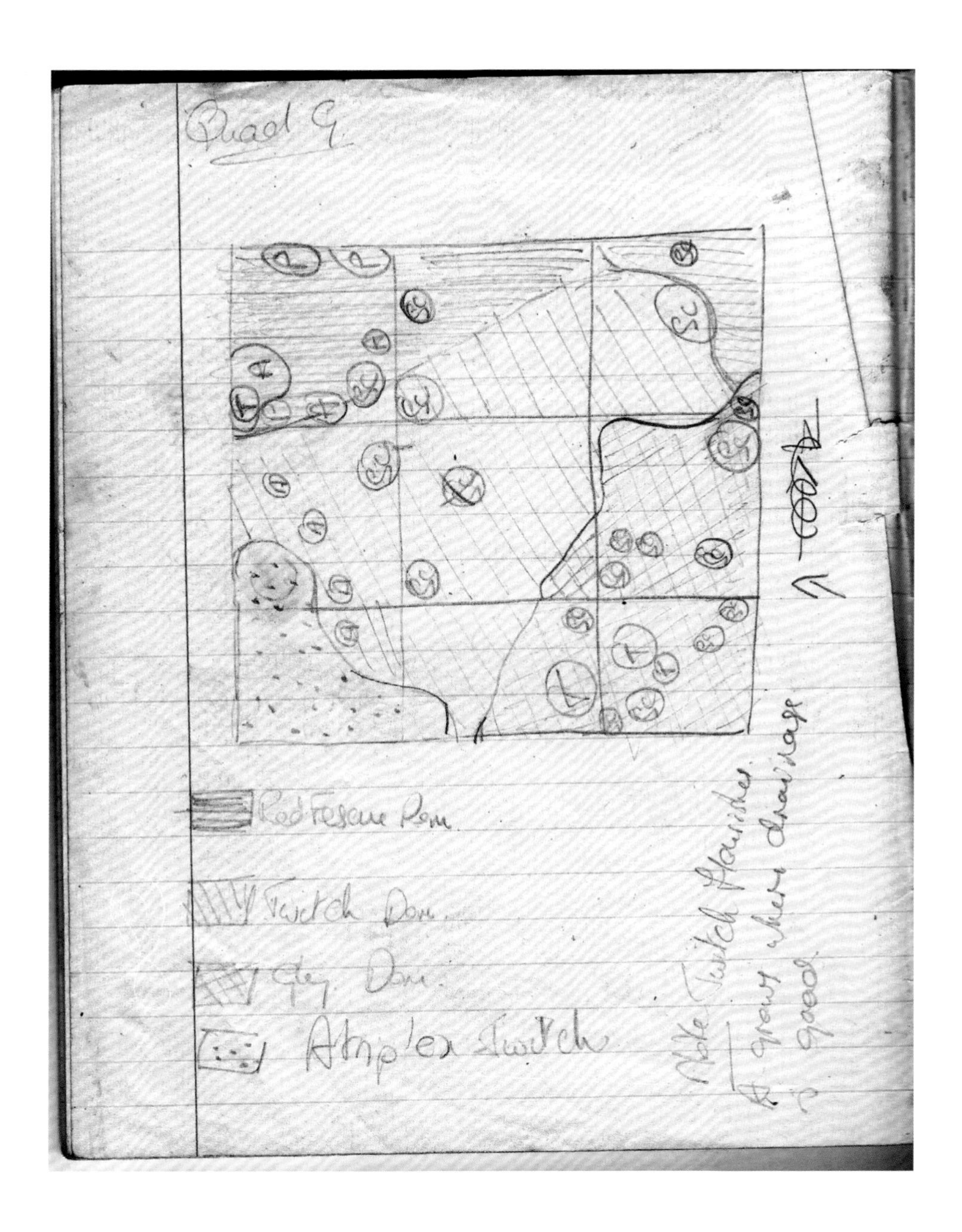

Sketches from Beth's old Colchester High School for Girls exercise book of salt marsh quadrats, done with guidance from Andrew Chatto, about 1940.

Salt marshes of the River Colne near Elmstead Market – a watercolour by Andrew Chatto.

Andrew scribbled notes for Beth on White Barn Farm headed paper reminding her of the importance, among many other things, of keeping a diary, listing all plants, making a general map of the district, and digging in different areas to find variations in subsoil. Beth's old red CHSG English notebook was soon turned into a rough book for the project, with pencilled plans and notes in a mixture of Beth's confident curly script with Andrew's scholarly smaller scrawl. In Quadrat J, he was excited to find *Enteremorphia* (sea lettuce), covered, he noted, under deep water at high tide. 'I hope', he wrote with the restrained excitement of a latent academic, 'we find *Salicornia* [glasswort or samphire] here.'

Occasionally, the scientific study gave way to more lyrical writing with inkblots and scratchings-out showing the passion of the moment. First from Beth:

> The driving mist of rain, the gleaming of the mud flats, the wailing call of a curlew, and the rising scream of gulls, the noise of raindrops pattering on our shoulders as standing, motionless, we watched in the seeming silence of desolation the rise and fall of a black-headed gull, the arresting scent of rain-washed spray-drenched marsh plants mingled with the ever-pungent salt smell of the sea; – misty blue-mauve of sea lavender, vast tracts of brown, green and grey – the reddish purple of young samphire, vivid grey green of marsh grass and streaming ribbons of seaweed stretched across the gently whispering mud.

Then, in Andrew's spidery writing:

> Dim curtains of rain rolling slowly across the flats like wraiths, heavy clouds, torn and shattered, mingling with the marshes in silvery mist. No sound in the wide greyness but gently hissing rain, broken now and again by a curlew's sad flute as he rises from his probing where the water laps. Far out across shining mud, the lovely, little river slips silently. Soft greens and greys and lavenders, subdued and melancholy spread over the wasteland. Water dropping and curling down the banks of innumerable creeks. A few swifts head up river, black beaks twisting through the rain.

Andrew Chatto, shy and already, at thirty-one, something of a social recluse, fell deeply in love with this beautiful, bright, feisty seventeen-year-old who so admired his botanical knowledge. It might seem an unlikely match given the big differences in their social backgrounds and the fourteen-year age gap. But, encouraged by their closeness over the salt marsh project, he repeatedly asked Beth to marry him. Her first reaction was to say no. Marriage was not part of her plan. She was determined to be a teacher and, because of the bar on married women in the professions at that time, marriage would have to wait. Also, for reasons that remain unclear, her father was against the marriage. Possibly he was concerned about the social divide, or maybe he just felt they were not suited. But Andrew persevered, clearly worried that Beth would disappear from his life once she graduated (or be snapped up by a younger, more outgoing suitor). His timing was good. Because of the war, there were few young men around and none, it appeared, with their shared interests and intellect. A year into her three-year teacher training course and halfway through the salt marsh project, Beth (and her father) relented and she and Andrew got engaged.

Soon, in one of the first indications of her impending move to a more middle-class way of life, Beth went to the Oscar Way studio in Colchester to sit for a photographic portrait, traditional among the middle classes and something she would continue to do regularly in the first years of her marriage. Formally posed, she looks older than her eighteen years. Photographed in profile, she stares wistfully into the distance, her straight hair softly permed as was the fashion. She wears a smart, full-skirted moss green jersey dress, a Chatto cameo brooch and, of course, her engagement ring.

By the time of the wedding, two years later in August 1943, just a week after Beth graduated from Hockerill, clothes rationing was becoming stricter. Although this was not a hasty wartime courtship, as so many were, there was no question of a white wedding dress. Each person was allocated just thirty-six ration points a year and with a wool dress taking up eleven points, no doubt there was some bargaining by the families to get coupons for her elegant 'dusty pink' gown and grey going away suit as described in the newspaper report. To the strains of the hymn 'Love divine, all loves excelling', with college friend Margaret as her bridesmaid, twenty-year-old Betty Little clutched her bouquet of red roses and exchanged vows with a relieved Andrew Chatto in the pretty little Saxon church of St Anne and St Laurence, Elmstead Market.

The future Mrs Beth Chatto, 1941.

2
From Betty to Beth

After the wedding, Beth did what many wartime brides had to do and moved in with her mother-in-law. Andrew lived with his parents in Braiswick since there were only labourours' cottagers on his fruit farm at Elmstead Market. His father had died from kidney problems in 1941, and Elisabeth Chatto, like her son, was shy, and also not in good health. Within a year of the marriage in 1943, she contracted flu and died.

'Weston', Beth's new home, was in the quiet, comfortable suburb of Braiswick, just west of central Colchester, seven miles from her family's modest police house and White Barn Farm in Elmstead Market. Situated down a small lane, it was an impressive home suitable for a prosperous middle-class family. Commissioned by Andrew's father in the early 1930s, it was a solid two-storey detached building, rendered on the ground floor and with overlapping tiles on the first. The front had something of the Tudorbethan style fashionable in the 1930s, with false timber framing, not at all typical of Essex or East Anglia but more reminiscent of the smart vernacular look of the Home Counties.

No doubt to Beth's great relief, the bar that forced women to leave teaching after they married had been temporarily suspended, 'pending the end of hostilities', allowing her to carry on in her chosen profession. A month after her wedding, she started her first teaching job, commuting by train through Colchester to Great Bentley, where the little red-brick Victorian primary school sat right opposite the station.

One of the first lessons Beth had to take was a sewing class. Material was in short supply because of the war and Beth found she had only six yards of green cotton for a class of forty-four Year Four children. An accomplished needlewoman, taught by her mother to do complicated stitching such as faggoting, Beth abandoned the

Newly-weds Mr and Mrs Andrew Chatto.

idea of trying to teach this class of eight-year-olds to do 'run-and-fell' seams and instead got them to bring in whatever unwanted materials they could from home. Old grey flannel trousers arrived, torn dresses and the like. From this literal ragbag of scraps, they made a collection of dressed toys – mainly elephants, given the grey flannel – the less adept shredding fabrics for stuffing. The toys were later put on display at a function in the village.

It was not long before the headmaster, Mr Swinden, realised that his new recruit was also an enthusiastic gardener. Beth was enrolled to help the teach the older boys the vital wartime skill of vegetable growing. An area of land behind the school was used to supply the school's kitchen. Some of the boys resented being taught gardening by a woman. 'My Dad don't do it like that, Miss,' one ten-year-old boy told Beth. 'Never mind,' she replied, 'there is usually more than one way of doing a job. You do it your way at home, and my way here, and we will see if there is any difference.' This was later to become a family saying in the Chatto home, that there was 'a Beth way of doing things and the wrong way of doing things'. It stayed with Beth for the rest of her life.

After two years, despite her love of teaching, Beth made the hard decision to leave the profession to help Andrew, who was struggling to cope on his fruit farm. Before the war, he had employed five or six men on the farm but half had been called up, leaving only the elderly. The arrival of half a dozen Land Army girls to replace the missing men was not the bonus it first appeared. Andrew's shyness proved a nightmare when it came to dealing with these forthright East End London girls. Beth stepped in. It was clear that someone needed to work with the Land Girls, as Andrew could not. She was deeply envious of their wonderful uniforms. Coming out of London these girls had no appropriate clothes for the country but

A fruit box label from Andrew Chatto's apple farm at Elmstead Market.

they were eventually provided with suitable breeches, skirts and waterproof coats. These seemed to Beth unattainably smart.

While the early Land Army girls were volunteers, by 1943 they had been conscripted and Beth soon realised they found the work very hard and were, in many cases, desperately lonely. She also thought them to be useless on the farm. 'They melted like snow, one after the other,' she later remembered. Beth had been taught how to prune the apple trees by Andrew, a principle that she believed could be applied to almost everything else that needs training in the garden. She now learned a great deal more, about handling people and poor soil. The farm was laid out to apple and pear trees. 'Cox's Orange Pippin', while always in demand, was tricky to grow, needing as much care as a pedigree animal and protection from every pest and disease.

Fruit farmers are at the mercy of the weather. Andrew's prediction was that out of five years, you would be lucky to have one really good year, three average, and one disastrous year when you lost a lot of money and had to borrow from the bank. Beth soon found that part of her role as his wife was to accompany him from the bank to the solicitor and then on to the accountant in those difficult years. While it may have been good experience for running her own business later in life, it left her with a horror at the thought of being short of money or, even worse, having to borrow.

Once the war was over, Beth's life began to follow the pattern of many middle-class English wives: a career given up on marriage and then full-time motherhood. With the men returned from war to work on the farm and the Land Girls gone, her help on the farm was no longer needed. There was to be no opportunity to go back to teaching when her first daughter, Diana, was born in March 1946, followed by Mary in September 1948, both delivered at home at the very dawn of the National Health Service. Beth had a difficult birth with Mary but she was supported by a live-in maternity nurse, Hilda, for three months and then by various mother's helps, always on hand to look after the small children.

Nevertheless, there were certain jobs Beth insisted on doing herself. There was always the feeling of having taken on somebody else's house and taste, in this

Frontage of Weston, the Chatto family home in the suburb of Braiswick, Colchester, with Beth and Andrew's young daughters, Diana and Mary, playing in the drive.

case, her late mother-in-law's. The decision was made to redecorate. Although the Chattos were financially comfortable, Beth, with a frugality learned from her parents, decided that she and Andrew would do it themselves, which they did. The cleaning of the great old gas stove in Weston's kitchen was another project. The latest model when Andrew's parents put it in the 1930s but showing its age by the 1950s, cleaning it took Beth the whole day. Its cast iron pipes and grills and enamel doors and shelves were all dismantled, violently scrubbed and put back together, with Beth repeating her mother's mantra, 'If a thing's worth doing, it's worth doing well.' Her later diaries are full of references to cupboards cleaned, curtains washed, items polished. She could never 'sit and rest in a mess'. During times of stress, be it from family or the business, Beth always found 'making my home clean and tidy is comforting.'

One day, she read an article in *Picture Post* magazine about an ingredient called Agene that, when combined with the white wheat flour used in shop-bought bread, caused an increased risk of heart disease. (It is no longer used.) She was horrified and from then on baked her own wholemeal bread. She was a great fan of Doris Grant, whose books *Your Daily Bread* and *Dear Housewives* became a cult success. (Many people saw Grant's recipes as 'cranky'. This led to the choice of name for the first health food restaurant in London – Cranks – which opened in 1961. Beth was a regular visitor when she was up in London.)

Grant was also a promoter of the Hay System of food combining. While Beth never following this regime religiously, she began a lifetime of careful eating which included making her own yoghurt with a lactic culture, warmed to just the right temperature on the stove. In later life, she was scathing about the 'discovery' of kefir cultures, saying she had been using them for many years. She later also became a vegetarian and kept a record of virtually all her meals in her diaries. However, she was never evangelical about vegetarianism and would usually eat whatever was put in front of her so as not to upset her host. Similarly, while she made a point of not drinking alcohol at home with the family, she never refused wine when away.

SUCH WAS BETH'S BOUNDLESS ENERGY that, despite the endless baking, cleaning and decorating, she managed to find time for more enjoyable home-centred interests such as flower arranging and gardening. The garden at Braiswick was typical of the 1930s. Neither of Andrew's parents was much interested in growing plants. When the house had been built, they had the garden laid in a conventional manner with a large lawn surrounded by herbaceous borders, a rose pergola leading to a rose garden and, beyond that, lavender hedges and a vegetable garden. They all backed on to open cultivated fields, typical of Essex's arable farmland. Andrew had added a rockery and a wild garden which he maintained himself.

When Beth arrived in 1943, she found the garden came with an ancient jobbing gardener. They were soon at loggerheads. His style was old-fashioned parks department, with spades polished and everything pruned into neat, round, tidy bushes; Beth's was not. One day she found him cutting all the flower buds off a lilac bush. She was furious, and astounded to think that he could not recognise the big fat buds as future flowers. Beth asked Andrew if she could do more in the garden. Andrew, who was always busy either with the farm or, more often, in his study with his books, wisely agreed. They shared the work in the main garden but Beth took over the vegetable garden, helped by farm workers to do the digging. In her second book, *The Damp Garden*, she described how she made her first 'and ever memorable' visit to the Chelsea Flower Show just after the end of the

Diana (left) and Mary (right) with Beth in the rose garden at Weston.

Second World War. Although plants were in short supply, Beth began to make plans for the garden.

The soil at Braiswick was chalky boulder clay, which meant there were many things, such as ericaceous plants, that Beth knew she could not grow. She was also prepared for the state of the soil when droughts came and wide cracks would appear. Her greatest excitement was getting control of the vegetable garden, having been used to helping her father in his, and eating home-grown all her life. Here the soil in winter was like 'cut liver, dark and solid with white spots of chalk'. But she was already a keen compost maker, collecting all the leaf litter and straw she could to fill the compost heaps for later improving the soil.

The Chatto family pose in the rock garden at Weston built by Andrew Chatto.

Colchester is the oldest Roman settlement in Britain and the houses in Bakers Lane were situated alongside the ancient remains of the wide defensive wall. By 1945, these were no longer visible, but covered in elm trees. The Chattos owned several acres of rough meadowland beyond the family garden which were left wild, and this gave a sense of freedom. One of Beth and Andrew's immediate neighbours was Pamela Underwood, who ran a carnation nursery with greenhouses on land bordering their house. They had become friends during the war when Mrs Underwood – as she was always known – had used some of their land for growing tomatoes. Appropriately, she had been provided with Italian prisoners of war to help her grow these Mediterranean fruits. During Beth's difficult birth with Mary, she was distracted by the strains of the prisoners singing 'O Sole Mio' in the garden below.

Pamela Underwood watched as the Chattos' garden gradually changed from Andrew's parents' conventional delphiniums and asters, struggling and dying in the dry Essex soil, to the grey foliage plants that Beth, on Andrew's advice, thought would have a better chance. Soon she started growing her own grey-leaved plants – artemisias, ballotas, santolinas – which all went well with her carnations, and began to sell them through her nursery, called Ramparts after its situation on top of the Roman wall. She became an important influence on the young housewife and mother.

NOW IN HER EARLY TWENTIES, Betty Chatto was increasingly embracing her new middle-class status. As well as gardening and all her other domestic duties, she was becoming interested in flower arranging. The house at Braiswick was always decorated with flowers from the garden with Beth using containers such as small wooden boxes, many of which she continued to use for arrangements throughout her life.

She was not alone in post-war Britain in becoming addicted to flower arranging. Mary Pope had founded the very first flower arranging club in the UK in Dorchester in Dorset in 1949 after seeing what a morale boost it was for women during and after the war. Like Beth, Pope was an Essex girl, part of the Lay & Wheeler wine merchant family in Colchester. Pamela Underwood was a family friend and was inspired by Pope to start a similar club in Colchester. With the war only recently over, it was seen as a blessing for women to have a creative outlet. As the mayor of Colchester wrote in the local *Gazette*, '[Flower arranging] can be helpful in the home and as a hobby, and a last line for the ladies to relieve the tension of everyday shopping.' (In her scrapbook, Beth vigorously underlined the last phrase in red.) With rationing in place until the early 1950s, even buying a yard of cloth for dressmaking could be tricky, whereas there was, Beth felt, inspiration to be found in the hedgerows, as well as in flower gardens, with containers more likely to be a teapot with a broken handle than an elegant vase.

Beth was by now an experienced flower arranger in her new home in Braiswick. Although she did not have a husband coming back from years away at war, like many women, she felt relief and a shift towards the 'pleasures' of domesticity. While the marriage bar for teachers and other civil servants had been fully abolished in 1945, there was no question of Beth going back to her previous job. Instead, she poured her creativity into her flower arranging.

There was a chest in the hall (it still remains in the family) on which she used to arrange dramatic displays. In spring, she brought out a large pewter platter on which she would create a miniature garden scene. A piece of wood would become a small tree, the support covered in moss with primroses and snowdrops coming out of it, all surrounded by a shallow pool of water like a picture of woodland. While Beth showed little sentimentality towards other family celebrations, decorating the house for Christmas was a time when the girls joined in and helped spray foliage for her arrangements.

Around this time, in September 1952, Beth made the remarkable leap from being an unknown Essex housewife to appearing on a daytime television programme called *Leisure and Pleasure* introduced by Jeanne Heal. Heal, who had pioneered the idea of an intellectually challenging yet entertaining afternoon programme for women, sandwiched ten minutes of Beth doing an autumn arrangement between a performance by an accomplished lutenist, Desmond Dupré, and an interview with literary novelist Marghanita Laski. For this Beth was paid by the BBC 10 guineas (more than the weekly average wage at that time) plus her return fare from Colchester, 15*s*.6*d*. Her local newspaper, the *Colchester Gazette*, picked up on the story, reporting that the appearance came about after Beth had done an arrangement at a London exhibition, 'delighting people . . . with the clever way she arranged such common – and neglected – things as seeding dock, oak sprigs and fennell [*sic*]'. 'Her fascinating hobby', they continued, 'has sprung from a love of gardening and wild flowers. She believes, and proves, that the wild flowers, and even the weeds, of the hedgerow can be arranged into bright autumn patterns. Seed heads, twigs and dried leaves all play their part in this delicate art.' Although it was a long time before Beth was to appear again on television, the tone and pattern of her life was now set.

She became fascinated by the art of Japanese flower arranging and found an American book, *The Complete Book of Flower Arrangement*, with a chapter on it, that was to influence her for the rest of her life. The idea that intrigued her the most was the 'asymmetrical triangle', which the Japanese consider to be the most perfect form of flower arrangement. In her copy of the book, Beth underlined in red the following passage: 'It represents Heaven, Man, and Earth, and is founded on the Confucian teaching that man identifies himself with both heaven and earth. Its fundamental idea is the suggestion, not only of the living plant, but its surroundings and the conditions under which it grows.' This idea of triangular proportion was to embed itself deeply into Beth's psyche and ultimately formed the basis of all her planting principles. At this time, it stimulated her love of experimentation with a wide range of plant material in her arrangements.

An important part of the structure of the flower club movement was the regular competitions. Beth soon made her mark on those at the Colchester Flower Club. She kept her many prize cards with their fulsome comments all her life in a small cardboard stationery box. 'The very simplicity of your material and the superb way you have arranged it can reap nothing but my compliments,' wrote an anonymous judge of Beth's arrangement of a 'Spring Tapestry' at a Colchester Flower Club competition. More compliments followed: A dried material arrangement was

A dried arrangement by Beth using seed heads, cones and leaves from the garden, 1950s.

'outstanding in every way'. Another judge praised her 'simplicity of decoration and simplicity of material . . . quite lovely. Thank you!' and yet another wrote, 'Perfection in miniature – *quite, quite* faultless.' Whether a large table arrangement of white roses or a posy bouquet of 'astrantia, tiny violas and dianthus in a vase of Victorian porcelain standing primly in a blue-lined crochet mat'.

Some comments were positively gushing: 'I have never seen anything more beautiful. Clever in extreme – I would be proud to meet you.' In June 1955, Beth's arrangement for a hall decoration won a medal for the Best Exhibit in the Colchester Show and caught the eye of the visiting Princess Royal, Princess Mary. Beth would have been less pleased with the third prize place she got for her three vases of flowering shrubs, although the soon-to-be high priestess of the floral world, Julia Clements, noted that Beth's offering in the 'modern arrangement' category showed 'strong originality' and another of her arrangements 'could not be more beautiful'. Clements became a lifelong friend.

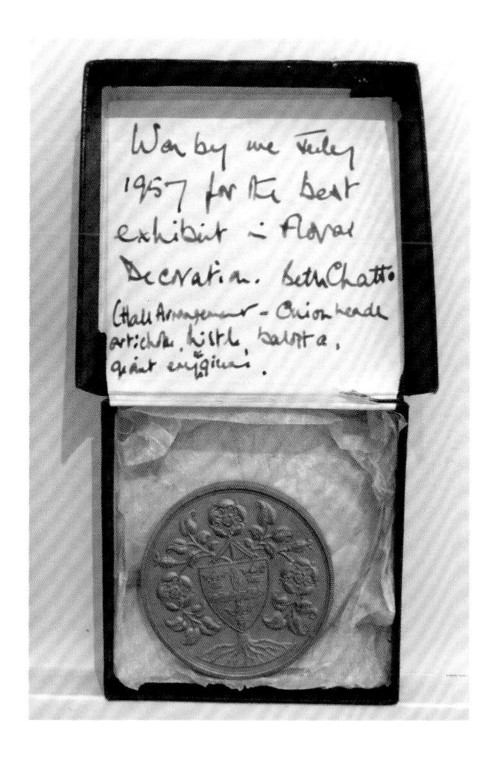

OPPOSITE Beth and her daughters, Diana and Mary, at the Colchester Rose Show in the mid-1950s.
LEFT 'Won by me July 1957 for the best exhibit in Floral Decoration – Beth Chatto. Hall Arrangement – Onion heads, artichoke, thistles, ballota, giant eryngiums.'

Not surprisingly, her competition success was also the start of Beth being in demand as a demonstrator. While she enjoyed being a member of the Colchester Flower Club, she later claimed to have been more reluctant to start appearing in public. However, in September 1953, Pamela Underwood telephoned to tell Beth that she had arranged for her to do a demonstration at the opening of another new club forty miles away in Saxmundham, Suffolk. Mrs Underwood, Beth remembered, dismissed any reluctance, saying firmly it was time for Beth to look beyond being a mother of two small daughters. Pamela Underwood was becoming known in the horticultural world as something of a martinet and Beth was left in no doubt that she had little choice. She 'counted to 10', cut flowers and foliage from her garden, and went on what was to be the first of thousands of demonstrations across the country. Andrew drove her – as she had yet to learn to drive – dropping her at the Saxmundham village hall before disappearing for a walk along the nearby coast. This pattern spurred Beth on to take her driving test. Andrew taught her, encouraging her to be sensitive with the clutch by putting an apple box (which she must not touch) in the middle of the drive at Weston. She passed her driving test first time. This did not stop her being nervous on her first evening outing when she had to take Pamela and her husband to a flower club meeting which involved reversing in the dark. 'Mercifully, I hit nothing,' she later remembered.

Beth was not aware that the woman who was starting the Saxmundham club, Elizabeth Colchester, had worked for Constance Spry. Spry had herself come from a humble background to become the leading flower 'decorator' – never using the terms flower arranger or florist – to high society. She also pioneered the use of material from the garden as opposed to the flower market, always encouraging her students to 'leave room for the butterflies' – unlike in the highly stylised arrangements that were becoming the norm from florists using stiff Dutch irises and strelitzias. Beth had met her when she came to demonstrate in Colchester and was very struck by her style.

Beth's own arrangements with hosta leaves, green hellebores and euphorbias were wildly popular. Shortly after her first demonstration she received a letter of praise from the chairman about her 'positively inspired demonstration . . . quite the loveliest I have heard and you so completely captivated everyone's hearts and those who were sceptical over our wild venture of a Flower Decoration Club I noticed were some of the first to come along with their year's subscription . . . I only hope you will continue the good work as you are such a born speaker and demonstrator – the two don't always go together.'

Her reputation as a lecturer and demonstrator was established almost overnight. A cutting from a local Suffolk newspaper recalled how it was Beth's appearance at the inaugural meeting of the Aldeburgh, Framlingham and Saxmundham Flower Decoration Club in October 1953 that was critical in the success of the club:

> A dull demonstrator could have killed the scheme from the start. Not everyone who can make a beautiful picture in flowers can also talk interestingly, and the good talker isn't necessarily first-class at flower decoration. So wasn't it fortunate that in Mrs Chatto those who were attending their first such meeting had the privilege of listening to and watching one who brings such intense enthusiasm to her task that she completely carries her audience with her as she talks of flower decoration while her hands swiftly and deftly create a masterpiece . . . The Club, now no longer a baby in arms, but a very thriving child with 141 members, owes much, I think, to its first demonstrator, who got it off on the right foot.

Beth's first visit to the Chelsea Flower Show as an exhibitor came soon after. In 1956, she was part of the Eastern England team who took part in the first ever Chelsea Flower Arrangement Show. She designed the stand for the group's contribution, 'Summer Profusion'. Having drawn every arrangement showing their shape and colour, Andrew made a 'beautiful' cardboard model, while Beth's twin brother, Seley, designed and painted Grecian columns for the stand. Violet Stevenson, writing in *Amateur Gardening*, described a centrepiece 'profuse yet as dainty as could be'. Although the arrangement was anonymous, the clues that this was Beth's arrangement were there: 'young larch, Solomon's Seal, *Arum italicum* leaves, wood spurge, guelder rose, berberis, tassels of sycamore'. It was, wrote Stevenson, 'a true gardener's arrangement'. But she also picked up on a change of direction in flower arranging. 'Gone the pie-dish and the few stiff leaves, the well-defined focal point.' (However, thirty years later, in 1984, Beth felt even more strongly that there was 'little originality or concern with the beauty of single flowers and leaves . . . all too much like floristry – perhaps because they choose flowers to *last*'.)

Gone too was Betty Chatto. It is unclear exactly when and why this happened. Beth and Andrew were always known as 'Bet' or 'Bessie' and 'Dan' among the older generation. Yet sometime early in her marriage, and certainly by the time of her first television appearance in 1952, the change had been made to the more middle-class 'Beth'. Was it suggested by Andrew? It is possible he thought it more poetic than Betty and sounded better alongside his surname. It may have been encouraged by Pamela Underwood, who, having been the first to spot Beth's talents, saw her as her protégée. But she never used her own first name publicly, always insisting on being addressed as 'Mrs Desmond Underwood'. While this was strictly correct according to the etiquette of the 1950s, it was thought old-fashioned and somewhat pretentious within the horticultural world. Whatever the reason – and Beth later claimed convenient amnesia over why she made the decision to change her name – the stage was set for her emergence into a new world of social, intellectual and horticultural society.

3
Benton End

One reason Beth was enjoying her flower arranging demonstrations so much was that it got her out of the house. While it had not been obvious during the war years, it was now clear that Andrew was naturally reclusive. He suffered with anxiety and hated socialising, while Beth – fourteen years his junior, pretty, vivacious and somewhat highly strung – loved it. Outside the home, she was making a mark on Colchester society and the Chattos were asked to many local events. One such was a fancy dress party given by Dr Robert 'Bob' Sauvan-Smith and his friend Peggy Kirkaldy. Sauvan-Smith was the family doctor and delivered both Diana and Mary. Kirkaldy was a socialite who held regular literary salons and was a close friend of novelists Jean Rhys and Dorothy Richardson. As usual, Andrew did not want to go but Beth was enthusiastic and unfazed by such intellectual company, seeing it as a rare chance to get out and meet people. She later complained in her diary about Andrew's reluctance to socialise: 'no television, no parties, no fun'.

Leaving Andrew at home with his pipe and his books, she went on her own, dressed as a Spanish señorita. Using the sewing skills she had learned as a child, she made her costume with leftovers in the Chatto family dressing-up box, including the late Mrs Chatto's old lace collars and cuffs and a black lace overskirt with a long train. A young Ronald Blythe, later famous as the author of *Akenfield*, who was working at the local library at the time, was also at the party, wearing a Tudor costume previously worn by his friend the artist Paul Nash in 1911. But it was Beth who won first prize. She was, Blythe remembers, 'very pretty'.

Andrew and Beth at Col de Sevi, Corsica, July 1951.

As soon as their daughters, Diana and Mary, were old enough to be left with Beth's parents, Andrew and Beth started to take holidays abroad. Whereas Andrew, after his early years in California and family holidays in Europe, was an experienced traveller, for Beth, who had never been abroad before, these trips were both romantic and exciting. Their first holiday was to Saas-Fee in Switzerland, now a popular tourist destination but in the early 1950s only just opening up to travellers after the war years and recovery. At that time, climbers with their guides were the only visitors during the summer, the fields full of fat dairy cows munching on grass and wild flowers.

The Chattos were accompanied on this holiday by two friends of Andrew's, Nigel and David Scott, also passionate plantspeople. The Scott brothers were the nephews of a close friend of Andrew's mother, Mary Pontice. She had started talking to Andrew's mother, a reclusive woman, while they were both waiting for the bus back from central Colchester to Braiswick. Within weeks, Miss Pontice, who was 'most definitely a lady', offered to come and live with this shy American after the death of her husband, becoming Mrs Chatto's companion and 'Aunt Mary' to the rest of the family. Nigel and David, because of their coincidental interest in plants, inevitably became friends with Andrew and then Beth. The quartet would later travel together to the Dolomites, Corsica and North Africa, at each destination observing natural plant associations as they varied at different altitudes. On this first holiday together, the group stayed at the family-run Pension Supersaxo. Beth remembered being served by village girls wearing the traditional local costume of black skirts and bodices with crisp white aprons and blouses with puff sleeves, their hair wrapped around their heads in neat long plaits, while the men strode around in lederhosen.

This was also Beth's first taste of seeing alpine plants in their natural setting. Being surrounded at Braiswick by the conventional British style of gardening of the 1930s, with its love of cultivars and all things improved by horticulturists – big peonies, delphiniums and roses – in these mountains her eyes were opened to a different world. Andrew showed her sedums and stonecrops, sempervivums,

Andrew with David Scott on holiday in the 1950s.

ABOVE Pen and ink drawing of an alpine forest by Andrew Chatto.
RIGHT *Fritillaria pyrenaica* – a watercolour by Andrew Chatto, 1954.

and dianthus all growing on the stony hillsides, water plants down in the gullies, trollius and polygonums in the marshy places. She began to understand why she was not able to grow these plants in the dry, chalky boulder clay of Braiswick. She realised that she was putting plants in the wrong place and trying to force them to grow. Plants, she began to believe, were like people. But the holiday was also an opportunity for new discoveries. 'We have one more big expedition to make,' she wrote on a postcard to her parents, 'then we shall potter about collecting odd plants. There are several we would like to bring but it will be a question of space!' Another card sent from close to Lake Lugano on a later holiday mentions seeing 'gentians, strawberries, thalictrum, campanulas, walnuts, chestnuts and many others you know. Am so longing to see you both and tell you.'

Fritillaria pyrenaica
First week of May 1956

She enjoyed being with the two entertaining Scott brothers. David was dashing and amusing, while Nigel was a handsome young man with a charismatic character. He had had 'an interesting war', volunteering in Norway and then joining a minesweeper which dropped him in the Mediterranean to infiltrate the locals, particularly in the French Alps. He was soon accepted by the shepherds and herdsmen and developed a passion for alpine flora. He also visited grand gardens on the Côte d'Azur and was quickly taken up by the rich and influential on the coast as well. All this was to stand him in good stead when he finally arrived back in Britain.

But when Nigel casually said one Sunday at Braiswick in the mid-1950s, 'Why don't we go over and see Cedric Morris?', having heard he had a notable garden, Beth, and to a lesser extent Andrew, were taken aback by the suggestion. They had never heard of Sir Cedric Morris and Beth's bourgeois instincts made her feel that this was not 'the done thing – just to go and invite yourself to see the gardens'. But a telephone call by Nigel Scott to Cedric was reassuring; he was happy to meet them and so they set off, winding through the Suffolk lanes full of cow parsley, none of them realising that this was going to be a day that changed all their lives.

Sir Cedric Morris lived with his friend and partner Arthur Lett-Haines,* at Benton End, near Hadleigh in Suffolk. There they ran the East Anglian School of Painting, with students such as Lucien Freud and Maggi Hambling. Their bohemian hospitality meant that Benton End became a melting pot of creativity. It was a place of both daunting simplicity and frightening sophistication. The house was so cold in winter that Cedric always used to disappear to the Mediterranean to find somewhere warm to paint. The plumbing was primitive and the water supply unreliable. When it failed, Cedric would go and dig a trench in the garden somewhere suitably tucked away and leave a pole with a flag at half-mast. When guests needed to go to the lavatory, off they would go and raise the flag to get some privacy. This arrangement was not just for the students but also for the stream of guests from London, the cream of London's artistic society, including poet Dylan Thomas, cookery writer Elizabeth David, as well as society flower arranger Constance Spry and botanical illustrator Mary Grierson. This was also where Beth first met plantsman Graham Stuart Thomas, in the garden talking to Cedric about music. Many of the friendships she made at Benton End, including those

Benton End, near Hadleigh, Suffolk, home of Cedric Morris and Arthur Lett-Haines.

with Thomas and Elizabeth David, influenced her for the rest of her life. Possibly in response to this new artistic society, Beth's Essex accent gradually began to change, according to those around her, to that of a confident middle-class woman.

Perhaps fortunately, Beth had no idea at this stage of the glamour of the world she was walking into – she was nervous enough as it was. She, Nigel and Andrew

arrived at Benton End that Sunday afternoon and walked across the gravel yard leading to the house with its Suffolk pink-washed timber-framed walls to find the back door. Like a stable door, it was in two halves and after knocking they entered into the large kitchen. Medieval in feel, it wasn't as big as a hall but it was high-ceilinged, with pink-washed walls and bunches of garlic and herbs hanging everywhere. It smelled, Beth remembered, of France.

On the walls were many of Cedric's flower and bird paintings, while more paintings were stacked haphazardly on the floor. About half a dozen people who had also been to see the garden that morning were sitting at the long, well-scrubbed table sipping tea from cracked mugs with Cedric, silk scarf knotted round his scrawny neck, at the head of the table wielding the teapot with his elegant long-fingered hands scarred by gardening.

Nigel, Andrew and Beth were warmly welcomed and invited to join the others at the table while the conversations, peppered with Latin botanical names, way above Beth's head at that time, carried on. After tea, Sir Cedric, tall yet elegant in his crumpled corduroys, led the three guests out and around the garden. Beth instantly felt like a child in a sweet shop. Whereas she struggled to grow a handful of fritillaries, here were carpets of hundreds of them that Morris had grown from seed, scattered himself and hand-weeded on his knees.

After this first visit, Andrew and Beth would go over to Benton End regularly. Although not a particularly practical gardener, Andrew was an ideal companion for Cedric, who was fascinated by his scholarly research into plant requirements. Initially Beth would walk behind the two men, chewing away on their 'wretched' pipes, feeling that she could understand only one word in ten of their scientific chat. But all the time, she was absorbing knowledge about plants – their names, behaviour and needs. More often than not, the Chattos would take their daughters, Diana and Mary, with them, trips the girls never forgot. While Beth, Andrew and Cedric wandered in the garden, Arthur Lett-Haines, or Lett as he was always known, would naughtily treat the girls to unheard-of luxuries such as a spoonful of caviar washed down with a Martini.

Sir Cedric Morris (1889–1982), with one of his paintings of Tall Bearded irises.

RIGHT Nigel Scott and Cedric Morris, holidaying in the Mediterranean in the 1950s.
OPPOSITE The garden at Benton End, where Morris grew his selected Tall Bearded irises.

For Nigel Scott, this was to be even more significant. It was the start not only of a long friendship between the Chattos and Cedric but also a relationship between Nigel and Cedric. Beth, despite feeling she was totally naive then, never forgot the moment of being there seeing these two men looking so good together, 'so right'. Nigel had taken a job at Scotts of Merriott Nursery in Somerset but the clocking on and off routine was not for him. A call from Morris asking him to Benton End rescued him from such routine.

Nigel began to work on the garden at Benton End, extensively clearing the land to increase the planting space. As the reputation of the garden grew locally, they began to open it to the public, bringing in yet more artists and plantspeople from across East Anglia. There was no formality in the garden, no contrived views or axes. The only framework was offered by walls close to the house. Other than that, it stretched out into a medley of old-fashioned roses and shrubs and an almost wild garden area. It had been criss-crossed with box hedging before Cedric arrived in 1940 but eventually these were dug out to allow more space for Cedric's much-loved bearded irises. In another part of the garden, a huge medlar tree cast deep shadows under which he grew shade-loving plants. 'The only rule was that there were no rules,' Morris said to a fellow East Anglian gardener, Jenny Robinson.

By this stage, Morris was amassing collections, particularly of the irises for which Benton End would become famous. He was looking for subtle combinations of colour, shape and texture in irises, with interesting veins, completely against the fashion of the time, which was for American irises in bright yellows and dark purples. He and Nigel did exhibit at the RHS shows in Westminster, which was quite a performance for them since they did not own a car. But, in general, Cedric had no time for the RHS or Chelsea, always happy to return to Suffolk and concentrate on

breeding his irises. He was generous with his plants and Beth regularly came home with something – alliums, fritillaries in variety and poppy seed carefully selected by Cedric for subtle colours.

Beth and Andrew also went on holidays and outings with Cedric, one year visiting Spain and Morocco together with Nigel. Beth sent a postcard to her parents enthusing about the plants they had seen: 'On the hills we have found crocus, iris, gladioli, lavender, fritillaries, tulips, narcissi, cistus and many more! Very few in flower – too early. We have to look for the leaves.'

In England, Cedric took Beth and Andrew to Sissinghurst Castle in Kent to meet Vita Sackville-West. After showing them round the garden, Vita took Beth inside the Tower and asked her to sign her visitors' book. Beth later wondered whether Vita perhaps felt that one day she would become a serious gardener. 'I don't know, but I was very touched.'

The more Beth went over to Benton End, the unhappier she became with the garden at Braiswick with its West Colchester chalky boulder clay. Morris came over occasionally to advise her but, one evening, Beth visited Benton End on her own for supper with Cedric and Nigel. After the meal, they were sitting and talking about planting as usual, when Cedric turned to Beth and said, 'You'll never make a garden where you are.' For Beth, it was a like a stone dropping, and her heart sank. She knew he was right. She would never make a serious garden as long as they were living in Braiswick.

4
White Barn House

ESSEX IS A COUNTY OF CONTRADICTIONS. Through Andrew's research, Beth had learned about the chalky boulder clay that surrounded her in Braiswick. As she was to write later in *The Damp Garden,* 'This whole area was formed about 10,000 years ago by the melting of the most recent glaciation (the Devensian), which covered most of the country about 18,000 years ago. During this last Ice Age, it had ground its way over the land to deposit, as it melted, a layer of clay many feet thick, incorporating nodules and grains of various rocks and chalk, from the hills it had passed over.' She knew too that she would never be able to grow the *Trollius europaeus* and *Persicaria bistorta* that she and Andrew had seen growing in the Alps in moist ground.

With Cedric Morris's words ringing in her ears, she struggled to find a solution. Her diary describes her lying awake at night, with her mind 'feverishly [leaping] from side to side – what can we do? What can we do?' Andrew had talked at one point of moving to Wales, where they had spent their honeymoon, but Beth despaired at that idea, knowing they would have nothing to live on. Instead, Beth proposed that the family should move from Bakers Lane to Elmstead Market, to the other side of Colchester, to live on what was still then Andrew's fruit farm. No matter that there was no suitable house there. What mattered was the site, the soil. There, encouraged by a comment made by the artist John Nash about its possibilities, Beth felt she would be able to create the garden she wanted. Andrew was not keen on the move. Neither – although they had little say in the matter – were the girls. Beth, on the other hand, was convinced that this would be the answer.

The farm had had more bad years than good throughout the 1950s. Although the British economy had recovered from the war – this being 1959, just after Prime Minister Harold Macmillan had told the British people that 'they had never had it so good' – farmers were now starting to struggle with foreign imports. Andrew was not a natural farmer and his heart was not really in it. In Beth's opinion, he did not go to the farm enough. Around this time, he was happier devoting time to drawing and painting landscapes and – not surprisingly – flowers.

In contrast, on a visit to their accountant, to which she always accompanied him, Beth's heart 'leapt with joy' when she heard him asking Andrew, 'Mr Chatto, have you never considered mucking your farm with your own boots?' The answer had to be a reluctant 'no'. He was of that generation of owner who gave his men their work to do and did not expect to join them. While Andrew Chatto never neglected the farm or its workers – he worked out their spraying programmes and taught them, men and women, how to prune – these were intellectual challenges he enjoyed and it was clear he preferred someone else to carry out his instructions. While Beth was happy to have helped during the war with the Land Girls, now she told Andrew firmly that it was not going to be her, firstly, because she was 'not devoted to running a fruit farm', and secondly, his men had never accepted her as anything other than the 'the guvnor's wife'.

The combination of the accountant's comment and Beth's intransigence weakened Andrew's resolve. When Beth heard that he was prepared to consider the move, she immediately started planning where on the farmland they might build their new home. The area she had her eye on was at the back end of the farm, an area of apparent wilderness and wasteland, unsuitable for growing fruit as it was either too dry on the gravel or too damp in the hollow. Although Beth could immediately see the possibilities, persuading Andrew was not so easy. He had deliberately kept this land untouched as a retreat for nightingales and other song birds. He encouraged local village children to come and explore, discovering dragonflies, butterflies and blackberries. Beth too loved the sound of the nightingales. They brought back memories of being woken up by her father and taken out into the garden with her brothers to a little copse to hear them singing. Although she knew that the birds needed to be undisturbed with brambly undergrowth during the nesting session, her priority was to find the best site for the house. Beth was also determined to make something of the spring-fed ditch which never dried out, even in a hard drought. Eventually Andrew was won over with the opportunity to make a water garden of some sort, the chance to grow many of the plants he coveted from their Alpine trips.

Andrew and Beth survey the site chosen for their new house on Andrew's farmland at Elmstead Market, 1959.

In 1957, a couple of years before Cedric's fateful comment at Benton End, Andrew and Beth had employed a young, newly qualified local architect called Bryan Thomas in an attempt to make Weston a more up-to-date and family-friendly house. An enthusiast for modernist open-plan living, Thomas had done away with the rarely used dining room and extended the hall into the study area complete with trendy room divider. Although they were not looking for a radical design for the new house, when the decision was made to start afresh on the site of the White Barn farmland in 1959, they went back to Thomas for ideas. The site chosen for the new house was on the south-western slope between the damp hollow and the higher-up deep gravel ground. This was not the cheapest option, given the challenge of building on a slope and with the change in soil structure, but it was the one Beth wanted.

ABOVE Local architect Bryan Thomas's design, described in *House & Garden* as 'a modern Essex farmhouse in an orchard setting'.
RIGHT In 1963, White Barn House was featured in *House & Garden* showing 'the differences in level mak[ing] each part of the open plan seem like a separate room'.

The site had originally been part of an ancient farm boundary. Before work could start, it had to be bulldozed to remove the tangle of brambles and blackthorn, wire netting and old fencing. Standing as sentinels on the north and south were two magnificent oak trees and a holly. Inspired by the sloping site, Bryan Thomas came back with a split-level design, dictated as much by the site as by his preference for modernism. The house, hardly altered today from Thomas's original plan, was featured in the design magazine *House & Garden* in 1963. (The issue also contained an article by the ever-influential Mrs Underwood on dianthus, a hint of how this may have come about.) The text shows Beth's influence on the layout: 'The architect's main task was . . . to design a combined living/dining-room and kitchen

where Mrs Chatto could discuss orchard matters with her husband, hear a daily record of events from her two schoolgirl daughters, and at the same time keep an eye on her baking.'

The article continues to describe the house, with its open-plan kitchen, small bedrooms 'extremely practically planned', and utility room, 'for discarding muddy shoes, flower-arranging, washing and home wine-making – one of the Chattos' hobbies.' Beth never allowed it to be called a bungalow – which, with its three internal levels, it was and is not. A few items of furniture came from Braiswick – the family chest in the hall, always a favourite of Beth's for a flower arrangement, a collection of opaque glass vases displayed on glass shelves above the bath in a style 'more likely to be found in Scandinavia'. A dining table and chairs by Ercol were brought from Weston (and are still in place today), while a pair of bamboo bucket garden chairs was acquired to provide an air of modernity if not much comfort.

Partly driven by economy and partly as a matter by taste, items were added gradually. Having lived through the privations of war, they had very much a 'waste not, want not' attitude. Curtains from the old house were cut down and reused in the bedrooms (and remain in one of the bedrooms to this day). Fabric for the sitting room curtains, in contrast, was commissioned to Beth's design and woven specially by a local firm just over the border in Nayland, Suffolk. Beth filled small notebooks with measurements and ideas. Much later, for the 'shady end' of the sitting room, she looked for '2 comfortable chairs preferably rounded backs' while she put 'Cedric's landscape above Aunt Charlotte's Chair'. There was no television in the house for many years but a battery-operated radiogram was an important feature for listening to music programmes and playing records.

Outside, the gentle slope of the roof was broken only by the dormer window of Andrew's study, accessed by an external stairway of floating steps which would never pass today's planning regulations. 'Mr Chatto', stated *House & Garden*, 'sometimes spends the evening writing: for although much of his spare time is devoted to gardening, he also finds time to write about the subject, and is at present working on a history of gardening plants', a project that would occupy him for the rest of his life.

This study would always be Andrew's refuge, a place to 'hide away' from his teenage daughters and the exuberance of Beth's enthusiasm, and to work on his research. Beth was constantly on the go in the house or garden with no time for relaxation, while Andrew retreated to his studies, often frustrating her with his lack of practical involvement and physical attention. Many years later, she confided to a friend that on one of her regular visits to Bob Sauvan-Smith, the family doctor, he suggested that Beth's various aches and pains were most likely a psychosomatic reaction to the lack of intimacy between her and Andrew. His suggestion that she might look for a lover deeply shocked Beth. She left the surgery with tranquillisers (often, in the fifties and sixties, known as 'mother's little helper', because of the regularity of their prescription to housewives).

The move was also not easy for Diana and Mary. They missed the old house, much as Beth felt she missed the house at Great Chesterford when she had to move as a twelve-year-old. At Braiswick, the girls had been close to the centre of town and their schoolfriends. That house had been large and rambling, whereas the new open-plan style left them with little privacy. Colchester was now a long bus ride away and the house seemed cold and bleak compared to the old-fashioned warmth of Weston. On the plus side, it brought the girls closer to their adored grandparents, Beth's mother and father. On William Little's retirement in February 1953, after nearly thirty-two years of service, they lost the police house in Elmstead Market that went with the job. Andrew bought them a house in Wivenhoe called Gardone where they lived for the rest of their lives, spending much time working on the garden.

It was with their grandparents that Diana and Mary had stayed when Andrew and Beth went on their plant-hunting holidays with the Scott brothers to the French Alps, Morocco, Corsica and the Dolomites. A rare family holiday with just the girls, in Scotland, Beth felt was not a success. Andrew had hired a Dormobile for the trip which included visiting Oban, Mallaig and Edinburgh, but it rained almost every day. Beth struggled to keep everyone comfortable and fed (and was not helped by the fact that she had left her 'pink pills' behind). With two teenage girls, emotions ran high and there were rebellious moods for all to cope with in a confined space, just as there were back at Elmstead Market.

Increasingly Beth's life focused on the garden. In moving to White Barn Farm, she finally had the variation of soils she had longed for. The Devensian glacier did not touch Elmstead Market but, as Beth wrote, 'the melting ice water did, gushing over the east coastal strip and laying down banks of gravel where the current was strongest, [and] sand and silt where it was quieter and still.' Across 12 acres of farm land, some of which was earmarked for her new garden, there were gradations from 'gravels and silt to the ancient London clay'.

Once the decision to move had been made, the daily visits to manage the farm became opportunities to bring precious plants to their new home. While, for the girls, there was the tedium of cold winter weekends spent tramping the muddy building site and playing Snap in the family Morris Minor Traveller to while away the time, for Beth it was the chance to fill the car with boxes of bulbs and perennials potted up, ready and waiting for their new homes. Storing the plants at White Barn Farm until they could be planted was a challenge, given the rampant rabbit population. Workers on the farm were inveigled into making a small compound protected by wire for Beth's 'treasures', which included her euphorbias, ballotas, lavenders, santolinas and thymes.

Although she was aware of the sensitivity of dismantling the old garden at Weston too much – and that mature shrubs were hard to move – there were some special plants that had to go with her. Bryan Thomas remembers being roped into helping Beth and Andrew load a precious ginkgo tree into the back of the small station wagon, while every bulb or rhizome given to her by Cedric Morris was dug up and transferred to Elmstead Market. While Andrew went to see to the farm, Beth would spend days digging up the bracken that overran the ground surrounding the house. Like bindweed, bracken has deep roots that regrow if they are snapped. It would take years before it was finally eradicated from the garden. But at this moment, with the walls of their new home growing every day, Beth enjoyed the work, relishing digging into the light, sandy soil rather than the heavy clay she had had to contend with in Braiswick.

The family moved in before the building was finished, with polythene covering the windows, no stairs and only one completed bedroom. Once there, it became obvious the new site had its own challenges. Although 1960 had been Britain's wettest summer since 1727, in July that year Beth picked up a handful of soil in the top garden and it trickled through her fingers 'like sand in an egg timer'. This was her first experience of the dry soil conditions typical of this corner of Essex. In a rare and brief moment of nervousness, she thought, 'My God, what have we taken on?'

5
'Unusual Plants'

While the article in *House & Garden* magazine had said that 'the garden is Mr Chatto's special hobby,' the accompanying photograph of Beth standing by the nascent pond area in boots and anorak suggests a different story.

Although selling Weston released money to build their new home, margins were tight and having lived through the war, both Andrew and Beth were careful with money. With the expense of the house, there was little left for the landscaping projects Beth had in mind once the builders left. Photographs taken then show a blank canvas surrounding the house, muddy and bleak, giving no hint of what was to come. With Beth's precious plants safely in place on the upper level of what was to become the nursery, they were able to take their time deciding about the paths and terraces around the house.

After a year of settling in, she and Andrew had made their plan, walking the paths, shuffling them in the bare soil. Determined not to have any straight lines in her borders, Beth made beds with gentle curves. A bulldozer was hired to level the soil so that the steps and terraces could be put in place. Beth had an article from a 1959 architectural magazine on 'Paving patterns and their uses' which gave her ideas for the design of the path leading up to the front door. In 1961, Andrew laid the slabs and cobbles, Beth having been inspired by a garden in the Hansa quarter of Berlin, with the random rectangles of precast slabs in pale grey and red matched with small granite setts.

In 1974, Beth wrote an article, again for *House & Garden*, describing the initial attempts at landscaping the site: 'We marked the few trees worth preserving and put a bulldozer through the 20-foot-high thickets of blackthorn, the bramble-tangles, the willow swamps and tall bracken beds. On the boundary, we left an acre or two undisturbed, where a badger still has his sett and nightingales nest in

ABOVE Beth and Andrew designed the shallow steps to accommodate the natural slope of the land.
RIGHT Early planting around the terrace close by the house.

the nettle-beds.' As yet Beth had no clear plan about what to do with the ponds. Meanwhile grass was sown and trees were planted. Like many owners of new gardens, they put in too many and over the years some had to be removed because of competition between the roots. Beth was now being helped by Winnie, a local woman who became her 'weed-killing queen', eventually controlling the bindweed that rampaged across the site. A concerted campaign to build up compost was helped by contributions of straw and mulch from the farm. A constant supply was needed to feed the new beds as they were marked out.

While the move to White Barn Farm was a joint project that kept both Andrew and Beth busy, it did not solve the problems in their marriage. Outwardly, it

ABOVE An early aerial view of the gardens and nursery, closely bordered by fruit farming.
RIGHT Beth in 1967, the year the nursery first opened to the public, beside the makeshift cold frames.

Harry Lambert was also much involved with the establishment of the ponds. These were made by damming the spring-fed ditch. Soil had to be taken away but without proper equipment they were able to achieve a depth of only about 4 feet. One morning, Beth woke to find the ponds had emptied and were now just puddles of water in a sea of mud. Water voles – 'Ratty' of *Wind in the Willows* – had bored through the dam and let the water flow down the stream to the farmland below. Elaborate piling using wooden logs, such as Beth had seen and admired at the Cambridge Botanic Garden, was out of the question. Once again, Harry came up

with a practical and economical solution, using a combination of concrete blocks and angle irons left over from anti-rabbit fencing.

Money was tight, and Beth was always looking for ways to make it go further. Seed trays were knocked together from spare wood and she used her local contacts to acquire plastic coffee cups from a nearby factory to use for cuttings. They were not ideal, as they let the light through and did not last long, but thousands were used in the early days of the nursery. One evening, the Elmstead Market Women's Institute came to see the garden. A few days later, one of the visitors, Madge Rowell, returned to seek out Beth, offering to help in the garden. They agreed a trial month, which was to turn into a twenty-five-year friendship until Madge's death in 1992. Her husband, Harold Rowell, made a little block to punch holes in the bottom of the cups for drainage. In addition, a young girl called Eileen, also from the farm but recommended because she always spent her lunch breaks with her head in a gardening magazine, came over to give Beth a hand for a month and stayed.

By now, Beth's relationship with Hans was part of all their lives. Hans's family was also aware of the situation. According to her diaries, seeing him daily 'recharged' her and, with Andrew's acceptance, they began to travel together to France and then to Holland to meet Hans's family. It became an accepted routine that Beth would see Hans almost every evening. On one of their regular meals out in Colchester, they bumped into Lett, who was with Cedric and Lett's housekeeper Milly. Despite her nerves about such a meeting, Beth wrote later, 'Lett was [very] kind. It was alright.'

While Beth wanted to support Andrew in his intellectual efforts – and to learn from him – she needed the romantic excitement that Hans offered her. Nothing highlighted this more than the birthday presents that she received from them each year. In 1968, Hans gave her 'a sweet blouse, a stunning green swimsuit and a bottle of "Noa Noa" [scent]'. From Andrew, she had '4 marvellous records – best of all Mahler's "Song of the Earth"'. The following evening, she and Andrew sat together listening to them – 'wonderful'. Throughout her life, music was extremely important to Beth. She and her brothers had had piano lessons as children but Beth had always found them rather a chore. While her younger brother David went on to become an accomplished organist and choir master, and Seley a music critic, Beth's interest was purely for relaxation. In addition to her growing record collection, she regularly listened to concerts on BBC Radio 3, particularly when she was in the kitchen. Her taste was for traditional classical composers, especially Mozart, Chopin and Beethoven, but she did, she told me, later teach herself to like Shostakovich while kneading bread.

In May 1968, on the spur of the moment, Beth asked Hans to take her to the Chelsea Flower Show. He bought her a yellow hat which, she wrote that evening, made her feel sixteen. But she wasn't so swept away as not to make some contacts and order a new arisaema. Later that summer, they went on a driving holiday around France but although she loved the countryside and the flowers, it also brought home their intellectual differences and the limitations of their relationship. 'Is there no man who can both love and inspire me?' she wrote despairingly in her diary in July 1968. 'I chafe because it is not the Great Thing.'

Inevitably, this emotional rollercoaster took its toll on both Beth and Andrew. In 1969, Andrew decided with little reluctance to sell the farm. It was a relief to them both. Finally, Beth believed, he would be able to concentrate on his plant research, retiring to his study in the sky and leaving Beth to run the nursery. '[The book] could be a great success . . . would be wonderful for us all.' But she also felt daunted at the thought of being a full-time businesswoman and the main wage-earner. 'Will I be able to make it succeed? Can I manage it well? I'm not sure.'

The sale of the farm gave them financial security but Beth no longer had access to its staff to help her out when needed. Plant sales helped towards Madge's and Eileen's wages and she also started to open up the garden more widely for visits. In December 1968, Beth, commissioned by the newspaper's editor, Jo Firman, began a regular series of articles for the *Essex County Standard*. This was an opportunity to publicise both the garden and the nursery. She was also thrilled to get a mention in *Amateur Gardening* in August 1969; this brought in lots of enquiries. One visitor asked if she designed gardens but Beth was reluctant to do so – 'could lose me money', she noted in her diary. Beth wrote to flower club secretaries telling them of her range of plants with green flowers and unusual foliage and seed heads, aimed at arrangers. It was the rarity of her plants at this point that made it an obvious choice to call the nursery Unusual Plants.

The first plant lists had been typed and, with finances always a consideration, nursery visitors were given blank labels and a pencil with which to write their own labels, a tradition that lasted well into the new millennium. But, in the 1960s, with

For many years, plant catalogues followed the same format with only subtle colour changes to the familiar cover.

the establishment of the formal nursery business, Beth paid out the substantial sum of £94 15*s*. 6*d*. for the catalogue design work that was to be her trademark. The first of her catalogues with their distinctive leaf design appeared late in 1967. The format was to remain unchanged for thirty years. Beth was much amused when, a few years later, Graham Stuart Thomas sent her a cutting about a modern ballet by the American choreographer Martha Graham, *Night Journey*, featuring Martha Graham carrying a stick of leaves inspired, he felt, by Beth's logo. She doubted it but was delighted that he thought so.

In 1969, the striking cover was a combination of dark green and lime- green leaves, with plants' prices ranging from 3*s*. to 15*s*. (the highest price being for a

Paeonia mlokosewitschii). The forty-page booklet divided into the lists that were also to become her trademark: Hot, Dry Conditions; Cool Conditions; Dry Shade; Waterside and Bog; Ground Covering Plants; Handsome Foliage Plants; Seed Pods; Grasses and Sedges; Heathers. When the first copies arrived from the printers, Beth was thrilled: 'drooled over my catalogue. It is nice!'

There were always new additions to the nursery, whether gifts or purchases from fellow gardeners such as Cedric Morris and Pippa Rakusen, director of Harlow Carr Botanical Gardens in Yorkshire (now run by the RHS), many from Alan Bloom's hardy perennial nursery at Bressingham in Norfolk, bulbs in their hundreds from Van Tubergen's and seeds from organisations such as the Alpine Garden Society. In the autumn of 1972, seeds of around seventy different plants were sown, from arums, aquilegias and anemones through to thalictrums and trollius. When given a selection of year-old RHS seeds by a friend, Beth simply soaked them, noting in her stock book that she 'then hoped for the best'. Treasures such as *Veronica telephiifolia* came from alpine specialist Will Ingwersen, along with *Uvularia grandiflora*, set to remain in the garden's catalogue as a firm favourite for a shady, humus-rich soil. Boxes would arrive from plantswoman Elizabeth Strangman's nursery in Kent.

Within a year of opening, the nursery was selling enough plants to cover its costs. But the conversion to decimal currency in 1971 and then the introduction of 10 per cent VAT in 1973 put a strain on the small business. Andrew no longer helped Beth with her catalogue, so she called Pamela Underwood to ask her advice on pricing. There was no question, she replied. Prices had to go up to cover the cost of VAT, just as hers had at Ramparts. So in 1974, Beth wrote in the new catalogue, 'It is with reluctance that I am obliged to ask you to make the following adjustments.' Plants previously priced at 20 pence went up by 5 pence and so on through the price range. Only the unpronounceable *Paeonia mlokosewitschii*, or 'Molly the Witch', remained a steady decimal equivalent of its pre-decimal price, now 75 pence. The 'bloody hard work' of putting together the catalogue was relieved by a steady stream of visitors. The artist John Nash encouraged Beth by bringing her a drawing of hellebores. Cedric Morris gave her some treasured double colchicums. Many of the plants given to her by Morris became the bedrock of the nursery. 'Today', she wrote in November 1972, 'I made 29 plants of Cedric's Dirty Pink Poppy.' (This became a nursery stalwart as *Papaver orientale* 'Cedric Morris'.)

The volatility of the relationship between Beth and Hans put additional strains on the business. Beth had an agreement with Hans that she could take on his female workers during the spring and summer months when they were not needed for picking or pruning at Park Farm. But this gave him a degree of power over her which led to resentment. There were also the regular boundary issues, mainly to do with Hans's use of the reservoir, that infuriated her. Falling outs would often ensue. There were also rumours of other women. While Beth knew she had no rights over Hans given her determination to stay with Andrew (although they no longer shared a bedroom), the strain took its toll and, in 1974, she had a nervous breakdown. There was no question of the nursery closing. With support from her daughters and the small band of staff, she was able to return to work after several weeks. But the anxieties remained and she was to continue taking tranquillisers on and off for many years.

At the end of 1974, Beth summed up her concerns, not just about the nursery but about the world situation as well.

> The proofs of my new catalogue lie on my table, I have no idea if the prices will cover me, or for how long. I think the strain of uncertainties, the endless tales of violence and horrors both manmade – such as the troubles in Ireland overspilling into bomb outrages here, or football hooliganism, or strikes and rebellions, to appalling natural disasters . . . all this makes for deep feelings of insecurity, a feeling of living from day to day. It cannot go on – we must eventually settle down, albeit to a different way of life – how different? How can we prepare ourselves for a lower standard of living? I am sure we must . . . I'm glad I deal in plants. They have a value outside money – they can't be cheapened and they have the gift of renewing my faith and hope. To walk into my greenhouse this week, this last week of the old year, has been a joy – so packed with plants, green and healthy, the smell of growing – warm and earthy, the new shoots, new roots, new promise for a New Year.

Beth inspecting the spring-fed ditch that supplies water to the gardens' pools.

By 1975, the catalogue of plants had grown to sixty-four pages. The leaves that year were emerald green and bright lime but remained faithful to the original design. In these early days of the nursery, Beth was prepared to try anything that would raise the profile of the garden. Rather reluctantly, she did agree to design and plant a garden near Woodbridge, Suffolk, one of the very few commissions she ever took on. She also did some planting at Le Talbooth restaurant near Dedham on the Essex/Suffolk border. She soon found out that it was too challenging to take on outside jobs while running the nursery. She was always generous with advice on what to plant, but her first and main love remained her own garden and its development.

Visiting flower arranging clubs, on the other hand, was something she was much happier doing, as she had been for twenty years. Between March and October 1973, she visited nearly fifty different clubs across East Anglia, keeping a detailed record of the plant material she took with her. She would pack her Volkswagen van with half a dozen or so vases, collections of plants, seed heads and foliage, together with boxes of plants to sell. On arriving at the hall, she made a point of going up to the group of women always there to prepare for the event, and ask them, 'Have any of you had a stomach operation?' More often than not, they would look at her with amazement. Then she would say, 'Which of you can come and help me cart my boxes?' Having had her appendix out the winter of the move from Braiswick, Beth felt she knew only too well the dangers of heavy lifting. Before the demonstration the plants hardly sold; but once Beth had done her arrangements with them invariably most of them went, and Beth would return with an empty van and money to help pay her wage bill.

Beth had never lost touch with her mentor and close neighbour at Braiswick Pamela Underwood, who was still running Ramparts Nursery specialising in silver-leaved plants. For years, Mrs Underwood exhibited at the fortnightly RHS shows and at Chelsea. Beth would occasionally help her and in 1966 took her daughter Diana along to help set up Mrs Underwood's stand in the Grand Marquee. As was standard practice, Mrs Underwood had previously set up a replica of her allotted space outlined in chalk on the floor of her barn. The position of plants in pots was decided and space allowed for flower arrangements, a particular speciality of hers given her long involvement with the flower arranging clubs. Known as the 'Silver Queen', she

Throughout the 1970s, Beth kept up a rigorous programme of flower arranging demonstrations to promote the garden and nursery.

was a daunting character, described by Beth as a 'slightly stooping figure, ash gently drifting down the front of her London suit from the inevitable cigarette held between her lips, her fine eyes half closed against the smoke'. By the early 1970s, she had won a sheaf of Banksian medals from the RHS for her exhibits and she was awarded the Veitch Memorial Medal 'for unfailing high standards'.

From the start of the nursery in 1967, Mrs Underwood had tried her utmost to persuade Beth to exhibit at the then-fortnightly RHS Halls shows in London. For a long time, Beth held out, feeling that the nursery wasn't ready. As Beth remembers, 'It's no good holding out a carrot and then snatching it away.' She didn't have stock available and knew it would take several years to build up. It did not help that Andrew discouraged her with warnings of humiliation, seemingly convinced that it was beyond her capabilities. However, she was encouraged by the comments of other leading plantsmen who were beginning to make their way to Elmstead Market. Valerie Finnis, soon to become a good friend, and her husband, Sir David Scott, visited in April 1973. 'She almost went off her head about the garden,' remembered Beth. 'Then said she'd never seen such good quality plants! All very exciting.'

By January 1975, Beth felt confident enough not just in herself but also in her stock to put her name forward for a stand at the late January 1975 RHS Show at their halls in Vincent Square, Westminster. 'Got up feeling strongly that I should be at the RHS hall! Rang Pamela. She was delighted to have hooked me at last! I wonder if it is the right thing for me?' Time was to prove it was, but at that moment Beth still had doubts. Although she had been to the famous annual Chelsea Flower Show several times, she had rarely been to any of these more intimate shows where visitors could not only see new plants but buy them as well and chat to the growers in a more relaxed atmosphere. Friendly they may have been, but it was still an honour to be allowed to exhibit there.

Beth's challenge was the choice of plants to take with her, the main criterion being that they would have to look attractive in the lean gardening month of January. Learning from Pamela Underwood, she did a mock-up of the stand which made her realise she would need more space. She called the RHS to request a space of 12 × 10 feet. But there were still anxieties. It looked good at the nursery but how would it look in the hall? There were ajugas with purple or metallic foliage, *Iris foetidissima* 'Citrina' with its orange-podded seed heads, and pink-tinted tellimas with purple backs to their leaves. However, it was the hellebores that Beth was most proud of. 'It does take several years to establish a plant so that it's got 40, 50, 60 flowers on it . . . *Helleborus foetidus,* the lovely pale green one, the Christmas Rose, *Helleborus niger,* [they] were out of this world.' They were all dug up and wrapped in peat and hessian to keep them alive.

Beth’s first show at the RHS Halls, Westminster, 28 January 1975: ‘We received a Silver Medal, to my delight and astonishment. In this picture I have just arrived in the hall, off the train, spattered with rain.’

The aim was to create a little winter woodland garden. A big branch of the twisty willow *Salix babylonica* var. *pekinensis* 'Tortuosa' was brought inside to force, making a little tree to cast some shade. On a low bench in the nursery packing shed, the plants were laid out around it, the aim being for it to look as though it was a planted border with no pots showing, as they were covered with peat. Crumpled-up newspaper was stuffed between the pots to cut down on the amount of peat needed. Her team at the nursery consisted of Harry, seven regular women workers and some casuals. But it was Madge Rowell who was chosen to go London with Beth that first time. 'If Madge hadn't been sitting beside me, I would never have arrived,' remembered Beth, 'because I had never driven to London before in my life. What a thrill it was to be driving up the Embankment and facing Big Ben. It was all such an [adventure].'

As ever, parking in central London was a nightmare with the constant threat of the car being towed away. But once inside the hall, the two women were welcomed by RHS staff who showed them where the trolleys and water were to be found. 'I think they were quite tickled to have two women because nearly all the other stands were being put up by men.' Madge bought lunch – 'we sat on our bark bags and ate pork pie, chips and bananas' – although Beth was too excited to eat. That evening, they drove home to Essex exhausted but pleased with their efforts.

The next day they returned to find the stand had been awarded an RHS Flora Silver Medal. Despite her long and close friendship with regular medal winner Pamela Underwood, Beth later claimed that she had had no idea that medals were awarded. Given that she had been helping Mrs Underwood regularly at Chelsea since the mid-1950s, this seems to be disingenuous. 'Immediately I had to be photographed for *Garden News*,' wrote Beth in her diary. 'Soon I was congratulated by Tony Venison, Christopher Lloyd, Graham Thomas, Lanning Roper, Valerie Finnis – it was overwhelming.'

It was not just the fact that they were women that caused curiosity. Two eminent plantsmen were heard arguing for hours over Beth's *Arum italicum*, as to what it should be called. What might have been more of a problem was that one of the judges, whose name remains a mystery, wanted to have Beth disqualified since he claimed all the plants, with the exception of some of the hellebores, were weeds. The argument was on what constituted a weed. Beth's aim was to introduce the public to the charms of uncultivated or species plants: that is, those produced naturally without any human intervention in their breeding. She had little interest in the cultivars they were surrounded by on the other stands. Hillier's were showing forced flowering cherry trees. Yet another stand was a sea of polyanthus, 'all shades of blotting paper . . . pinks and yellows [with flowers] the size of pennies', so far removed from the delicate primroses that had been dug up from White Barn Farm.

When Beth and Madge returned in February, there was yet another query over the medal their stand was awarded. No question of disqualification this time – rather the Silver Flora Medal they had won was by a subsequent RHS Council meeting raised to the higher level of a Silver-Gilt Banksian Medal. The judges were particularly impressed with the hellebores, bergenias, golden feverfew, purple-leaved bugle and the double primrose 'Marie Crousse'. What they didn't know was that Beth had run out of plants the previous day during staging and, on her return home, had had to call on Hans to drive her back up to London in the evening with more stock.

At the Great Autumn Show in Vincent Square, the success continued with yet another Silver-Gilt award. Once again Hans – this time with his son, Hans Junior – helped bring the plants up to London from Elmstead Market. Beth was furious to discover that one of them – she wasn't sure which – had bent a miscanthus grass to make more room for sitting. The next day, Christopher Lloyd came over to tell her she had won the Silver-Gilt Flora Medal. While the Gold Medal still proved elusive – Graham Thomas told Beth they had only just missed out on it – the stand was the only one to be featured twice in the show coverage in the RHS's journal later that year. 'A big thrill for me,' she wrote. 'Imagine it! A year ago, no, I wouldn't have dared.' Now featuring plants for a drier soil, it was the *Eryngium agavifolium*, *Salvia officinalis* 'Purpurascens', kniphofias, colchicums and 'the spiky seed heads of *Allium albopilosum*' that caught the eye on her 'imaginative exhibit'. To crown a winning year, *Kniphofia* 'Little Maid', raised by Beth, a short red hot poker with delicately graded lemon flowers, was awarded a Certificate of Preliminary Commendation at the show and was selected unanimously for trial at Wisley. And, more significantly, she was invited to show at Chelsea the following spring.

6
Chelsea Gold

JUST AFTER CHRISTMAS 1976, a large envelope addressed to Beth dropped through the letterbox at White Barn House. Inside was a folded sheet from the RHS showing the layout of the Great Marquee, where she was to exhibit for the first time. There, in the midst of this vast arena, was Beth's first stand looking, she felt, 'much like a postage stamp'. Even more daunting was the surrounding company. Next to her stand was a company that could not have been more different in style. Thomas Rochford specialised in house plants and was known for its spectacular displays orchestrated by Betty Rochford – formal and exotic, a breathtaking show of colourful house plants. In contrast, Beth was determined that her display would look like her garden in mid-May.

Rather than just taking plants in pots from the nursery, Beth worked busily through January and February, lifting mature plants from the garden. She sketched out a design on a piece of paper. The square stand was to be divided diagonally, showing plants and grasses for cool conditions in one triangle and plants and grasses for dry sunny conditions in the other. Running from corner to corner were the large dividing plants: a gunnera and a big ornamental rheum. These had been lifted and sheltered in a polytunnel, but nothing was forced by heat.

On the 'dry' side were alliums and galtonias still in bud. The look Beth was aiming for was a microcosm of the garden at White Barn House. She was not concerned that some things were still in bud, just doing what she liked the look of. Several years later she remembers one of the international bulb firms losing a Gold Medal because the night before, when the judging was done, not everything was fully out. It was no consolation to the tearful company men to find they had all opened the next morning. The judging tradition had to be maintained.

A plan sketched out by Beth with ideas for her first stand at Chelsea in 1976.

Although Beth had been pressured by some of the show committee to exhibit, once on site there was no help to be had from any of the officials. On their first setting-up day, she and Madge arrived without a trolley or even watering cans,

CHELSEA FLOWER SHOW 1976

BETH CHATTO

Plants and Grasses for Cool Conditions

Ajuga 'Burgundy Glow'	40p
Ajuga atropurpurea	35p
Alchemilla hookeriana	40p
Angelica archangelica	35p
Arum italicum pictum	50p
Aruncus sylvester	40p
Carex buchanani	45p
Carex morrowii 'aurea'	65p
Epimedium 'sulphureum'	45p
Euphorbia palustris	40p
Euphorbia robbiae	40p
Filipendula 'variegata'	40p
Fragaria vesca 'variegata'	35p
Helleborus corsicus	60p
Holcus mollis 'variegatus'	40p
Hosta crispula	50p
Hosta Fortunei 'aurea'	50p
Hosta Fortunei albo picta	50p
Hosta sieboldiana	50p
Iris Foetidissima	40p
Iris japonica variegata	£1.00
Lamium 'Beacon Silver'	50p
Lamium 'aureum'	40p
Lunaria biennis variegata	30p
Luzula maxima 'marginata'	40p
Lysimachia 'aurea'	35p
Milium effusum aureum	40p
Miscanthus japonicus	50p
Miscanthus japonicus 'Zebrinus'	75p
Petasites japonicus	50p
Phormium montanum	85p
Phormium tenax	85p
Polygonatum multiflorum	40p
Polygonum bistorta 'superba'	40p
Polygonum carneum	40p
Pulmonaria saccharata	40p
Pulmonaria argentea	40p
Rheum palmatum 'rubrum'	£1.50
Saxifraga umbrosa 'variegata'	35p
Scrophularia nodosa 'variegata'	60p
Smilacina racemosa	£1.00
Tiarella cordifolia	35p
Trifolium repens 'purpureum'	35p
Trollius europaeus	40p
Vinca minor 'alba aurea variegata'	40p
Viola labradorica	35p

Plants and Grasses for Dry Sunny Areas

Acaena adscandens	35p
Allium aflatunense	35p
Alchemilla hookeriana	40p
Anthemis cupariana	35p
Arabis Ferdinandi coburgii 'variegatus'	40p
Artemisia arborescens	50p
Asphodelus cerasiferus	50p
Asphodeline lutea	40p
Ballota acetabulosa	40p
Ballota pseudodictamnus	40p
Crambe cordifolia	50p
Cynara cardunculus	50p
Dactylis glomerata 'variegata'	40p
Eryngium agavifolium	40p
Eryngium variifolium	40p
Euphorbia cyparissias	35p
Euphorbia polychroma	40p
Eriophyllum lanatum	40p
Foeniculum vulgare 'Bronze'	40p
Festuca amethystina	40p
Helichrysum angustifolium nanum	50p
Iris pallida 'variegata'	£1.00
Lychnis viscaria	40p
Lychnis viscaria alba	40p
Marrubium cylleneum	50p
Melissa 'aurea'	35p
Origanum 'aureum'	35p
Ruta graveolens 'variegata'	50p
Salvia argentea	40p
Salvia officinalis 'icterina'	40p
Salvia officinalis 'purpurascens'	40p
Santolina incana	40p
Sedum kamtschaticum variegatum	40p
Sedum middendorfianum	35p
Sedum murale	35p
Sedum roseum	50p
Sedum spathulifolium	35p
Sedum spathulifolium capa blanca	35p
Senecio leucostachys	40p
Senecio 'White Diamond'	40p
Thymus 'Doone Valley'	35p
Thymus 'Golden King'	40p
Tropaeolum polyphyllum	£1.50

The plant sales list for Beth's first Chelsea Flower Show, showing the divisions of recommended conditions.

having struggled to find a parking spot for their van. Their allotted area was marked out with four stakes in the ground. But there was a camaraderie among the exhibitors. She was quickly loaned a watering can by the Dutchman Romke van de Kaa, later head gardener at Great Dixter, and the Rochford group and Hillier's nearby offered help. All were to become firm friends.

The stand took shape in line with Beth's earlier sketch. The cool condition plants were arranged around a shallow pool with wood blocks giving the effect of shaded stepping stones. In contrast, the dry sunny plants were displayed with pebbles and paving. The plant list, now professionally printed rather than the typed sheets she had used for the RHS shows in Vincent Square, was of two columns, each of nearly fifty plants for the different conditions. After the days of setting up, with Madge by her side, they arrived on the opening day of the show to find they had won a Silver-Gilt Medal. She later admitted privately to being disappointed in not being awarded a Gold. But there were compensations. 'From the minute we stepped on our stand we were besieged – many people delighted because we were "their kind of gardening".'

In that first year, Beth's was not the only stand showing herbaceous perennials. The big difference was in her style of showing. The majority of other growers staged their plants in serried ranks of visible pots at various levels. The other big difference was that these displays of delphiniums and lupins were all cultivars, always a dirty word in Beth's vocabulary. While they were aiming for size and colour, the fashionable style of gardening for so long, to Beth this was anathema.

One of her plants was *Lamium* 'Beacon Silver', whose 'quiet grey-green leaves' caught the eye of Anne Scott-James when she wrote about the 1976 Chelsea show for the RHS's journal later that year. 'It was nice to greet the first appearance at Chelsea of Beth Chatto, whose plants I have long admired in the gardens of connoisseurs,' Scott-James wrote. 'Mrs Chatto groups her plants not only to get variety of contrast and shape; she says it is even more important to associate plants which enjoy a similar habitat. In other words, a hosta could never look right alongside a helichrysum, and she finds it sad that hostas are so often planted in sunny places. Among her groups of plants for cool, leaf-mouldy spots was *Arum italicum* 'Marmoratum', an 18-inch beauty with glossy green spear-shaped leaves veined in cream, shown off by Solomon's seal, hostas and ferns.' For some, the arum has become a garden pest but it remained a favourite foliage leaf for Beth when flower arranging.

It was an article in *The Sunday Times* by Graham Rose three weeks later that Beth credited as the turning point for both the garden and the nursery. Rose had been among the throng of journalists who had congregated around Beth's stand on the Press Day at Chelsea that year. Lanning Roper, Christopher Lloyd and Tony Venison were also there but it was, Beth remembered, 'The *Sunday Times* man [Graham Rose, who] bided his time, then tried hard to make himself my press representative'. When most of them had gone, Rose remained, '[a] tall lanky figure propped up against one of the tent poles wearing his typical navy blazer and the cigarette dangling out of his mouth'. Symbolically waving the other journalists away, he whispered conspiratorially to Beth, 'Don't you take any notice of them, I'm going to look after you.' 'Oh dear,' thought Beth of this flattery, and put it down to the fun of the day.

But Graham Rose kept his word and called her immediately she got home to arrange a visit to the garden. While Beth was keen to show him the water garden, it was the Mediterranean garden that interested Rose most. After an exceptionally dry winter and spring, the summer of 1976 was turning out to be the hottest in Britain since records began in 1880. Many parts of the country had seen no rain for months. Beth's Mediterranean garden was, by her own admission, indeed looking like the Mediterranean, 'a bit parched'.

Rose saw an opportunity to show his readers what could be done even in these exceptionally dry conditions. He asked her to draw out two planting plans for a medium bed and a small dry bed together with the respective plant lists. Not a problem, thought Beth, until she realised that he wanted her to do the plans then and there. 'This is for *The Sunday Times*, it's important . . . I need to think about it,' she retorted. But Rose reassured her that any of a thousand permutations she could do would be acceptable and that she could do it. 'It was the first lesson . . . that he taught me, that you don't always have time to take a month to do something, you've got to do it now because people wanted it yesterday.'

The pair took a sheet of paper and began making notes, Rose smoking with ash dropping on the table. By 10 o'clock in the evening the plans were complete. The next Sunday, the half-page article, entitled 'Blooming arid', appeared. Beth's dry garden, Rose wrote, was full of plants scoured especially from southern Europe, 'whose resistance to drought had been tempered by natural selection and adaption . . . All those in the drier south and east of England, who, armed with hosepipes, have been battling last summer and this to save their wilting gardens will avoid both anxiety and expense if they follow either of Beth Chatto's plans on this page.'

The article, with its detailed recommendations of over thirty drought-loving plants, was to turn 'Unusual Plants' from, in Beth's own words, a 'really rather parochial little business' into a nationally known name. Up until that time, orders were usually for £10 or less and from local addresses. It was clear that the combination of Chelsea and Graham Rose's article had given the business a huge national boost. As Seley, Beth's twin brother, pointed out after a career in advertising himself, 'I don't think you realise the thousands of pounds' worth of publicity you've been given.'

Less than a week after the article came out, Beth had four publishers asking her to write a book. 'Shall I?' she wrote tentatively in her diary. She also had the RHS wanting her to join a judging committee but she turned that down, stating she had no time. They persevered and tried again to get her not just on to a committee but the RHS Council as well. Yet again, she refused.

The Great Autumn Show 1976 brought another Flora Silver-Gilt. She later heard that one committee had voted unanimously for a Gold but a second committee had awarded the Silver-Gilt. 'Never mind,' she wrote that evening. 'One day I will do a really super stand and we will all think it's worth a Gold.' She did not have long to wait. While nothing can be quite the same as entering the Great Marquee for the first time on Tuesday morning to find you've won a medal, for Beth her second year, 1977, brought the added joy and glory of her first Chelsea Gold.

Putting on hold her hopes for a winter holiday with Hans, Beth had started making her plans in early January, at the same time as she started writing her first book, *The Dry Garden*. The set-up in May involved the usual early starts and anxieties. 'My stand is 21' × 18'. I could do with 25' × 20',' Beth wrote on the first evening. She was excited by her water garden side, 'but I dare not hope for a gold.' The next day, she and her new helper, Mel, finished it off. 'The Dry Garden went better than I'd imagined considering the plants were not first class. The addition of

The 'dry' side of the stand that won Beth her first Gold Medal at the Chelsea Flower Show in May 1977.

my Spanish wine jar was a touch of genius I think!!! Anyway we made up a nice focal point with *Gladiolus tristis* and variegated sisyrinchiums to repeat the pale colour.'

The next day, before the medals were announced, Beth was introduced to the Queen, even more than usually the centre of attention in her Jubilee year. She showed interest in *Alchemilla conjuncta*, described by Beth as 'a charming little plant [whose] small exquisitely cut leaves are backed with shining silk which forms a silver edge on the top side'. This brought Beth a lot of interest, not just in Britain

Beth poses holding her first Chelsea Gold Medal card, in May 1977.

but across the world from Holland to Australia, after photographs of her meeting the Queen appeared in the press.

Although the cover of the RHS journal on that year's show featured a rock garden of Victorian proportions, it was Beth's style of gardening that was causing ripples throughout the horticultural world. Visitors crowded around the stand. Beth was grateful for the security of the denim money pouch strapped round her waist, into which she tucked cheques and cash. Helen Dillon, who was to settle in Dublin where she created a famous garden, kept a note of the plants she ordered from Beth that year, together with Beth's words of advice: *Brunnera* 'Langtrees' (Had they bred it? Don't let it seed); *Crepis incana*; *Diplarrena moraea*; *Chrysanthemum* 'White Bonnet'; *Daphne odora*; *Kniphofia* 'Little Maid'.

Unknown to Beth, the RHS had made the decision, for reasons of 'austerity', to discontinue giving prize-winners actual gold medals. However, a 'retrospective' exception was made by the recently elected Secretary of the RHS, John Cowell, to ensure that Beth would receive a gold medal 'free of charge'. Beth wrote that she was particularly delighted to have it in the Queen's Jubilee year. September brought another Gold Medal at the RHS's Great Autumn Show. Magazines clamoured to interview Beth, with articles appearing in *Country Life*, *Harpers & Queen* and more by Graham Rose in *The Sunday Times*. The year was crowned by the wedding of her younger daughter, Mary, at the old church in Elmstead, decorated by Beth, as was the marquee on the White Barn House lawn, where guests danced into the evening despite the November chill.

Money was no longer quite such a worry. What concerned Beth now was getting the work done in the garden. 'I need a good manager,' she wrote. Over the next nine years, Beth was to win nine more Gold Medals at Chelsea. Her stand in 1978, she felt, was better than the previous year's, although 'already I am thinking of improvements for next year.' In September, she managed to have a holiday, this time without either Andrew or Hans. She flew for the first time by herself to stay for two weeks with two old friends from the Colchester Flower Club days, Jo Stewart and Pamela Walpole, who now lived in Cape Town. This naturally included a visit to the Botanic Garden at Kirstenbosch. Beth was impressed with the indigenous planting and longed to have had more time there.

In 1979, it was her stand, rather than a dated rock garden, that was featured on the front cover of the RHS's *The Garden* magazine. The only glimpse of colour in the photograph was the back of Rosemary Verey's red coat framed by the green and grey foliage of Beth's sumptuous stand. Richard Bisgrove, reporting for the RHS on that year's show, said that Beth's arrival, along with that of seed collector

Jim Archibald, 'heralded a new era in plantsmanship'. Because another exhibitor dropped out, she was offered a larger stand with an extra 10 feet, giving what Beth thought was a difficult area of 13 × 40 feet. Back in February, over a hundred plants had been repotted and kept in the new plastic tunnel inspired by a visit to Alan Bloom's nursery at Bressingham.

Beth was having to get used to being a celebrity. Lord Aberconway, then President of the RHS, tried yet again to persuade her to join the Council. Yet again, she refused, privately thinking that not only could she not cope with it but also she wasn't sure whether she would enjoy it. She knew she never would be a committee person.

Plants for Chelsea were protected during the winter months in a polytunnel and cared for by Beth and her team until ready for display in late May.

Beth's stand was visited by the Queen and other members of the royal family several times over the years.

Anglia Television came to film her with old hands Geoffrey Smith and Clay Jones. The garden, she remembered in her diary, 'was strewn with cables like entrails everywhere'. When the programme was broadcast a few days later, Beth and Andrew went to a neighbour's house to watch it in colour. Although colour television had been widely available since the late 1960s, it had taken long enough for Andrew and Beth to get any sort of television; there was no question of such an extravagant purchase while the old black and white set was still working. It was the popularity of snooker that had hurried the demand for colour television – and this was a sport that Beth later adored. 'Wretched Alex Higgins affects me unbearably – he's such a bundle of nerves,' she wrote about a championship match in 1985, 'while [Steve] Davis is brilliant and cold. I admire him but cannot feel warmth for him. The score, 7-0, to Davis.'

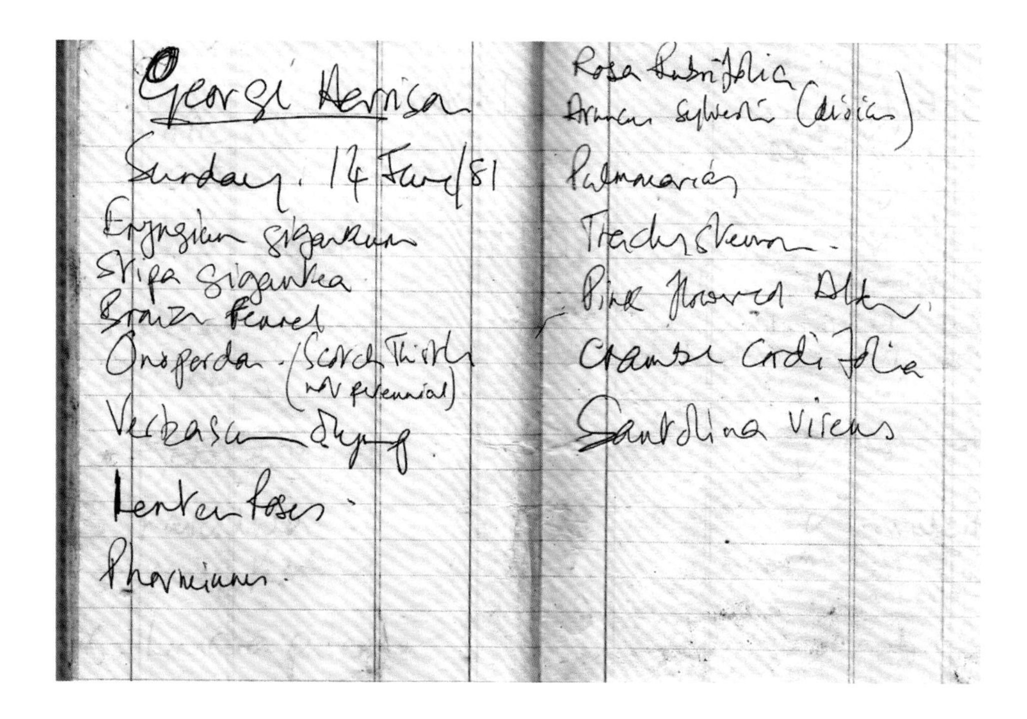
George Harrison
Sunday. 14 June/81
Eryngium giganteum
Stipa gigantea
Bronze fennel
Onopordon (Scotch Thistle) (not perennial)
Verbascum
Lenten roses
Phormiums
Rosa rubrifolia
Aruncus sylvester (dioicus)
Pulmonaria
Trachystemon
Pink flowered
Crambe cordifolia
Santolina virens

Notes made by Beth on a visit to the garden of George Harrison in June 1981.

Back at Chelsea, Beth had yet again been selected to be presented to the royal visitors. Dressed in what was now becoming her lucky suit, first worn for her daughter Mary's wedding, she met the Duchess of Gloucester, Princess Michael of Kent and Princess Anne. For the second time, Lord Aberconway presented her to the Queen. 'I always feel concerned for her,' wrote Beth that evening. 'I think she finds it all rather tiring but she said my stand was "very pretty"!!' It was pretty enough to get another Gold. The next day, she wrote in her diary, 'I was vastly relieved to see it this morning. My God, what an effort it is. I do not mean that I have to strain to do it, but rather I am driven from within by tremendous pressures to do always my best.' Throughout that 1980 Chelsea week, people crowded round her stand, notebooks in hand, desperate to get their orders in before she sold out. 'Don't come in the morning,' Beth would tell friends. 'Come late afternoon, when the crowds are getting less.'

One evening that year, she was asked to stay behind after the show officially closed at 8 p.m. Her mystery visitor was Beatle George Harrison. He wanted to buy the stand just as it was. More than that, he wanted Beth to become his gardener, developing his 36-acre estate at Friar Park near Henley-on-Thames, Oxfordshire. With regrets and excuses about her own garden, nursery and staff, Beth declined his offer. However, she did visit him and gave him suggestions, including *Eryngium giganteum*, *Stipa gigantea*, bronze fennel, *Crambe cordifolia* and the Scotch thistle, *Onopordum acanthium*, with the warning that it was not perennial. Beth's visit, it was later said by Dan Pearson, 'proved key as a confidence-building exercise. With typical practicality she had said: "You know, George, if you had an old sofa in your house that you didn't like you'd throw it out!"' Beth also introduced George and Olivia Harrison to grasses. Such became George's enthusiasm for horticulture that he dedicated his autobiography, *I, Me, Mine*, 'To gardeners everywhere'.

In future years, Beth was allocated extra space at Chelsea and was now able to divide the stand into three with displays of plants for sun, for shade or part-shade, and for damp conditions. The build-up and the Chelsea week itself were always challenging, with everyone involved running on adrenaline during show time. Beth and her helpers booked into a modest hotel, the Terstan in Nevern Square, Earls Court, recommended by Hillier's and used by many of the other stand-holders. It was to become their regular resting spot when a night in London was needed.

One evening, returning exhausted from the show, Beth opened the bedroom door to find a dead mouse in the middle of the floor. Undaunted, she picked it up by its tail and took it downstairs to show the manager, saying that she deserved a drink at the very least. Only later in the bar did she realise that her friends from Hillier's had played a trick on her and planted the mouse in the room. Despite this, they remained friends. There was a deep camaraderie between the surrounding stand-holders, with everyone ready to help each other out. While Beth, as a country girl, was not fazed by a dead mouse, she was still an innocent in other ways. On her first night at the hotel, she left her keys in her van, quite unaware of the insalubrious reputation of Earls Court at the time. The manager noticed and rescued the van before it disappeared into the night.

FOR TEN YEARS, the Chelsea Flower Show dominated Beth's life for many months of every year. It brought her world fame, publishing contracts, media coverage and world travel. But it also brought exhaustion and anxiety. She found 1981 a particularly frustrating year. Arriving one morning during the setting up, she was horrified to find that Notcutts, the famous Suffolk nursery exhibiting behind her, had put out billboards partially blocking views to her display. Furious, she had to rearrange her stand-making – she later told Charles Notcutt – 'the opposite side the main attraction and . . . the dark side [darker presumably due to the billboards!] to accommodate plants for a narrower, shady border, which fitted in quite conveniently.' She must have made her displeasure clear to the Notcutts staff because a few days later, she sent a rare apologetic letter to Charles Notcutt denying that she had been 'offended' by his exhibit. She finished the letter with the somewhat ambivalent words: 'I hope that you and your staff were rewarded by plenty of orders for your stand, which aroused a great deal of interest and admiration from the public.'

Just a week later, she fired off a letter to John Cowell, Secretary to the RHS, complaining that, while she had abided by their rules of not being allowed to sell books from the stands, she was surrounded by others flagrantly disobeying this, leaving her 'feeling foolish'. The paperback edition of *The Dry Garden* had just been published and Beth was cross at this missed opportunity for sales.

After Chelsea in 1981, her brother Seley, always her loyal supporter, sent her Brutus's famous lines from Shakespeare's Julius Caesar, beautifully written in a calligraphic hand to cheer her up.

> There is a tide in the affairs of men
> Which, taken at the flood, leads on to fortune;
> Omitted, all the voyage of their life
> Is bound in shallows and in miseries.
> On such a full sea are we now afloat,
> And we must take the current when it serves,
> Or lose our ventures.

The Baron does not want Yellow

Plants for Mouton.

The Grey Border

Asphodelus albus

Anaphalis trip.

" yedoensis

Anthemis Cupaniana

Artemisia discolor

" Valerie

" Powis

" purshiana

Ballota

~~Calam~~ Catananche

Centaurea pulcherrima

" pulchra major

Cerast. Columnae

TRAVEL NOTES 1

A trip to Château Mouton Rothschild (1981)

ONE RESULT of Beth's increasing fame was the number of invitations she received from the rich and famous, and the not so rich and famous, either to lecture or to give advice. Having successfully seen off Beatle George Harrison's offer of a full-time job, she then faced an invitation that was harder to refuse. In May 1981, at the end of a long day at Chelsea, Natasha Spender,* wife of the poet Stephen Spender, came to look at Beth's stand accompanied by a short man with grey wispy hair down to his shoulders and large animated eyes. He was introduced as Baron Philippe de Rothschild.

By now, Beth was used to famous names visiting her at Chelsea. Andrew was more taken aback when the Baron telephoned shortly afterwards to arrange a visit to Elmstead. This led to an invitation to Château Mouton Rothschild, fifty kilometres north of Bordeaux, which, to Andrew's dismay, Beth accepted. Baron Philippe (1902–1988), from the banking family, had led a playboy life in his youth, both as a racing driver and as a film and theatre producer. He then went on to run the family vineyards, turning them into some of the most successful in the world.

In these extracts from the first of her detailed travel notebooks, Beth describes her feelings of delight and terror at staying in such grand surroundings for the first time in her life, together with her nervousness of meeting the other famous guests.

A page from the small notebook Beth kept during her visit to Château Mouton Rothschild in 1981.

The Bordeaux Adventure begins. We are taxiing out now, my heart beats faster but I am determined to enjoy it and I do as the great 'bird' lifts of the ground, and the patchwork of England lies beneath me. Suddenly I see Windsor Castle, looking quite small, like a child's toy, and the gardens around it looking non-existent except for a kind of parterre. I spent the flight, once we crossed the Channel, marking David Hockney's plant and shrub list, dreaming of my new reservoir beds.

The journey was very short, only about 1 hour 20 minutes. Again my heart was beating slightly fast as I walked into the building, but there at the top of the stairs was the chauffeur with a large sign 'Mouton'. I laughed with relief, my baggage soon appeared and documents for my plants were not required. We drove for half an hour through dry looking country, scrubby pines, then the vines began. It was very warm, much warmer than at home. We had arrived at Château Mouton.

Natasha was waiting, but first, two servants came and took all my luggage, including my handbag! Then I met Baron Philippe heavily engaged with Joan Littlewood* in a great argument about the design of a low wall joining a gateway. We had lunch at Petit Mouton, on a small high terrace at the top of an outside staircase shaded by parasols with a marvellous room behind us full of needlework pictures and *objets d'art*. When I reached my room – my suite! – I found all my clothes unpacked and put away.

How shall I describe my room? My room *est chambre Italienne*. It is very simple, very beautiful, spacious, shuttered window to the floor look onto the vineyard. Shaded by a lime tree. No curtains. The bed I think French Empire? Gilded and curved, top and bottom. Voile-like pillowcase, scallop-edged, hem-stitched linen sheets turned down onto a white cover, flower sprigged. There are easy chairs, a desk with everything I need. A bow-fronted marble topped cabinet houses my small clothes. There are begonias in pots and a low table beneath a dark medieval looking painting. Next door is my bathroom with everything white, towels, robe, all so personal.

After lunch Natasha Spender took me round to view the garden and to discuss future plans. Compared with gardens in England, there is little established, but the Baron has planted good trees over the last 35 years to form a fine park. There are few plants and many of those are planned to go! Given time there could be some interesting planting, but already there are beautiful areas combining architecture, raked gravel and evergreens which could not be improved. To me they seem timeless and complete.

Two elegant French ladies arrived, guests for the night. Neither young but both handsome and chic as I have only ever seen French women to be. Dinner was at an oval table, clothed to the floor, candles lit with shades over the flames. We ate crepes served separately with noodles and such a tender meat in very thin slices, *gratinée* and served with a sauce. Cheese followed with a superlative sweet, *Figues à la Baronne* – sweet figs, brandy, lots of cream decorated with caramelised biscuits – all served with special wines. Finally, the golden and sweet d'Yquem. How do I come to be in this fantastic world?!! It is the most beautiful house I have ever stayed in.

How I wish I could understand more of the language. It is beautiful to listen to, and I catch some of it but I can only exchange such a little. But they have been very kind and friendly towards me – so now in my room I can say it has been an extraordinary day but I have enjoyed it every minute.

Thursday 25 September

Everywhere everything is beautiful. The buildings, walls, roofs, the raked and washed gravel, old established trees mixed, some evergreen, ilex and cedar, some deciduous limes sycamores and planes – the décor of the rooms – all white walls but marvellous floors, beautiful tiles, apricot coloured, sometimes plain, different sizes – one room had a squared pattern of polished wood with small apricot/brown tiles set between. Antiques – beautiful OLD things everywhere, chairs, table, pictures, lamps, *objets d'art*.

Morning started with a tray brought to my bed. I had woken late because the shutters kept out the light – the outside shutters were closed. Then I went with Stephen and Natasha to see the museum. It has to be seen, cannot be described. Such exquisite things perfectly staged, we only saw a fraction when Raoul – who is in charge of the wine – who chooses what we shall drink – collected us to see the cellars. We saw one section where there are 100,000 bottles. We saw from the huge polished vats through the cellars of barrels to the private collection of the house – some more than 100 years old, cobwebbed and still, having lain somewhere there through the horrors of several wars. All lit by tiny candelabras which looked like candles while Raoul carried a taper. Everywhere was perfection, not a sign of dust or dirt, not a dead leaf or stick.

Natasha and I spent some time considering the improvements that could be made. There alone can be I be critical, but no one, it seems to me, has had as much knowledge and understanding of gardening as they have here of interiors and viniculture – with the latter, I am impressed by the landscapes that they have created, the juxtapositioning buildings – the red curved tiled roofs and the horizontal planes of the vines – the odd groupings of trees, limes and cypresses – and the spaces between with the sky above. Outside my room as I sit now, this still après-midi, after a most elegant lunch – I am shaded by 5 or 6 limes which rustle in the breeze. They cast shadow onto my open shutters and the brushed gravel beneath them. Earlier this morning I lay in bed and listened to the swishing of the broom and saw the blue trousers which are the uniforms of the vine workers. Immediately beyond the limes stretch the level vines, hundreds of acres, the purple grapes not quite ready for the harvest.

The two most elegant ladies are still here. Both are works of art – they must have been very beautiful when young, but still are very good to look at. Their figures are very trim, exquisitely dressed and groomed. Their English is non-existent which encourages me to try my schoolgirl French. Before lunch we had drinks under the trees just below Petit Mouton, & Natasha showed the photographs I had brought of my garden and Chelsea. Immediately the two ladies expressed warmth and admiration – called me a great artiste (!!) – discovering as it were for themselves, although Natasha had introduced me as 'fameuse en Angleterre'. Stephen joined us and Baron Philippe until the garçon wearing his white jacket and gloves, announced that lunch was 'prêt'.

All afternoon till 6.30 p.m. we worked on ideas for the garden – not too easy! Monsieur has very decided views – will not have anything with yellow flowers. Philippe and M. Hirtz, the young gardener, came too. I was glad to retire to my cool room at 6.30 p.m. It is almost as tiring to earn my keep this way as working hard all day at home. Trying to absorb so much to understand some of what is being said, as well as trying to give something in return. I do not really feel that I can give very much. Certainly not in just a few days. But the Baron is wise, he knows it will take time, even though he is himself eager and impatient.

Friday 26 September

It is warm here and humid. I sleep easily, but wake several times feeling damp. I collected some damp peat and compost and planted the colchicums in front of the shrubs in the Petit Mouton. I'm not sure of them there, think they will need more sun but at least people can see if they like them. I potted up the cyclamen I brought. More guests for lunch, the architect and press officer (another handsome lady!) from Paris.

I really should learn French. I feel very uncultivated. I enjoy listening, understand a little but cannot take part. Later Natasha and I walked the Grey Border. I don't really approve of it, but we can only try to improve it.

I had the luxury yesterday of seeing some of the English people who were on the same plane as me being ever so slightly grand walking around here as tourists, while Natasha and I walked through the big gateway into the house. Silly me, I didn't feel grand only amused that I should be here! I'm not sure that I shall be able to do anything really interesting, there are too many restrictions – but, given time and patience, a lot could be done.

Saturday 27 September

I met Joan Littlewood after breakfast. I had been making a little plan for planting round her 'office' buildings and the walk to the swimming pool. There are thin straggly irises everywhere, all lovely shady places which could be full of ferns and hostas! Joan suggested I went with her to the market at Pauillac – I was delighted so we set off with her 'spotted Dick' of a dog called 'Tati', who hasn't the slightest road sense. Mercifully we spent some time going through rough land and scrubby lanes. Nowhere was really attractive, mostly a case of neglect or desertion. Presently we saw the wide curve of the river but we turned off into the town to see the last half hour of the market.

Then we drank a glass of white wine in a little open-fronted café facing the river. Joan is excellent company. She talks without taking a breath, but it is all interesting, I only needed to ask the odd question to set off another chapter of her highly involved life, her work with the theatre in particular. She is not gossipy or name dropping, cares very much – but is amusing, gets very agitated with Baron Philippe but rightly so about the gate [an ongoing design issue]. He knows it. But there is much shouting, stick waving, arms high, feet going round in circles. Joan sometimes wears a pink turban looking very 1940s. We returned to Petit Mouton to find Natasha and Stephen with iced drinks under the trees. We were joined by Philippe and his son-in-law Jacques [Sereys], who is an actor. All were delighted with the colchicums which have erected themselves.

Superb lunch – perfect melon and smoked ham in little rolls, cheese, then strawberries in strawberry syrup – or was it figs in white wine?!!! I have kept most of the menu cards. After lunch Joan took me over to Petit Mouton. Very Victorian but lived in by Philippine,* the daughter, when she comes from Paris. It is enchanting I think. Every wall covered with pictures, just the right furniture and treasures piled everywhere – but the bedrooms were my favourites.

By 4.30 p.m., Philippe and Jacques, Natasha and I were busy discussing new plans – to make lakes (ponds I would call them) at a corner of the park. I was not enthusiastic, I commented on the damage done to trees because roots cannot breathe, leaves falling in and fouling, etc., etc. The old man has sudden ideas and rushes into them without sufficient consideration. I hope they can all delay the ponds. They will add nothing unless they are made on a large scale and that means destroying too many fine trees.

Natasha and I spent the rest of the day staking and labelling the Grey Walk for replanting. We came in and added up the bill. But I am not terribly thrilled about the border. I don't think it is appropriate there, but it could be made to look tidier. There are lovely books in the library (thousands!) but I found several excellent garden books, one huge one, illustrated by Gertrude Jekyll. I love the smell as you walk up the curving snail-shaped staircase to the library and grand salon – of stone, wood and books. It is all very spacious but lived in, with cosy chairs and lamps always lit in the library. The dining table moves around, always a different cloth, very long, draped under the table, which is oval.

Jeannine is the name of the girl who looks after me, brings breakfast with a fresh flower every morning, with palest apricot tinted napkins, hand hemstitched and edged with white lace. The toasted croissant and coffee, I love.

Sunday 28 September

I woke early at about 7 a.m., I got up thinking to go for a walk with Joan but presently it rained. I opened my shutters to listen and smell the rain. I read Gertrude Jekyll's huge illustrated tome, *Garden Ornament*, till Jeannine brought my last breakfast at Mouton. She offered to pack my valises, but I preferred to do it myself. I was rather depressed with the Jekyll book. The utter grandeur – although the trees and shrubs were important, the architecture was dominant. I looked to see when the book was published, 1918. There were still very rich people, with plenty of gardeners. How many of the gardens illustrated would still be cared for. Few I guess. I would love to be able to plant once in my life a garden where I could associate plants in a really splendid setting.

12 p.m. – I went to the Baron's room to say my thank yous and farewell. I was embraced (!) and sat in a chair whilst he made a little speech, in effect that he found me *sympathique* and wished me to go on with the garden, that he 'counted on me'. I tried to explain that I had too many responsibilities but this seemed to add fuel to his argument that I 'worked' – that he liked people who worked, etc., more embraces. Thank goodness he is almost 80 years old.

7
Friendships and books

DESPITE HER BROTHER'S URGING, Beth did not exhibit at Chelsea the following year, 1982. She needed a break for a number of reasons. In January, she was told by her doctor that she needed to have a hysterectomy. There is never a good time to schedule such an operation but it had to be done that spring. 'This means cancelling Chelsea,' she wrote after her visit to the doctor. 'Perhaps it will be for the best,' she added in late January, 'a way to stop the ever-increasing feeling of tension and drive. The staff were all very kind and reassuring when I went to tell them after lunch.'

Not long after that decision, sadly but not unexpectedly, her old friend Sir Cedric Morris died in February aged ninety-six. He had become increasingly frail and Beth was relieved that he was no longer suffering. 'I feel thankful, and numb. [It is] impossible to say what he has meant to me. [He] influenced me in so many ways, made my life. And many, many others. A great man, and a very modest man, in his own way.' She had also written 'shy' in her diary but with her constant quest for accuracy, had gone back afterwards and crossed it out.

Later, just a photograph would remind Beth how much she missed Cedric, remembering 'the humorous yet searching look in his eyes, the sensitive mouth and long lean body . . . I miss him still [and] would so like to share plants again with him – and ridiculous giggles over nonsense too.' In 1990, with some reluctance, she revisited Hadleigh for the unveiling of a plaque in memory of Morris. She saw many old friends, she later noted, but felt 'cold and sad inside to find only the husk

Andrew Chatto and Cedric Morris, characteristically talking about plants with pipes in hand, about 1980.

Beth and Cedric Morris during a visit to the gardens at Elmstead.

of Benton End'. She described her final walk around the garden, now reduced to overgrown trees and shrubs, with here and there brave cyclamen and colchicums. 'I collected seed of Cedric's special umbellifer standing tall and spare, like himself. I thanked him and drove sadly home.'

If Beth's thirty-six-year friendship with Sir Cedric Morris had been one of the most influential in her life – she called him her 'spiritual father' – then her friendship with Christopher Lloyd, which had begun in the mid-1970s, was the best known. As she described in their joint book of letters, *Dear Friend and Gardener*, first published in 1998, she was given a copy of Christopher's bestselling book

The Well-Tempered Garden. She remembered devouring it, not stopping until she had read all the way through. She immediately felt she wanted to meet this person. While she instantly recognised a fellow plantsman from his wise words, she was furious and somewhat curious about his distaste for bergenias. To her mind, they were one of the essentials in her garden, her 'full stops', whereas to Lloyd, they were 'coarse and heavy . . . lifeless and depressing'. Beth wrote to him at Great Dixter and a postcard came back, with just the words 'come to lunch'.

Beth was unfamiliar with the south coast but had a flower arranging demonstration coming up in Brighton, so she decided to combine the two visits. There was an inauspicious start when her windscreen smashed in the tunnel under the Thames but she made it safely to her venue for the night, albeit coated in 'nuggets of glass', as she noted in her diary.

The next day, she drove over to Great Dixter and wandered around the empty nursery, coming upon the low shed door of the selling area with its 'Duck or Grouse' warning sign. While she was giggling to herself about that, the shed door opened and out strode Lloyd or 'Christo' as he soon became to Beth. She recorded the visit in her diary for 20 March 1974: 'CL looks a cross between David Kossoff (I think) and Peter F [a friend of Beth's]. He was charming and easy to talk to. Fascinating garden. The huge yew walls making outdoor rooms complementing the buildings. Beautiful. Very generous – gave me super plants. I enjoyed every minute. Delicious Italian aperitif. Red wine, a Burgundy with lunch. Mmm . . . Super day.'

In September that year, Lloyd made his first visit to Beth's garden. 'I thoroughly enjoyed having him,' she wrote later that evening in her diary. 'He was so friendly and easy. Lunch with Andrew was most pleasant . . . After lunch, went round nursery and collected him a few plants. By tea time, I was exhausted. Went to bed 6.30 p.m.' Five days later, she received a thank you letter from Lloyd which, she noted, 'quite bowled me over'. When she recovered, she spent all day on an article and plans for Michael Hoog of Van Tubergen's bulb nursery. 'I'm very slow,' she wrote ruefully. 'But I enjoy writing when it sounds good. Would practice make it flow like Christopher's?' Shortly after, she retired to her room with '2 ripe Williams pears to eat with my bedtime book – C. Lloyd of course!'

The one person Beth could not talk about in front of Christopher Lloyd was Graham Stuart Thomas, whom she had first met at Benton End. Fond of them both,

Beth soon realised that despite their common interest in plants, their personalities were as different as chalk and cheese: Thomas with his scholarly, introverted personality the chalk, while Lloyd, ever the outgoing, confident bon viveur, surely the cheese. It took Beth, she felt, far longer to earn Thomas's respect and affection than Lloyd's, yet it was Thomas's books to which she turned the most often. 'He would remind me to smell things, as well as look into the heart of them, or touch them,' she told me. 'I don't know whether Christo had that tactile, or sensual relationship with plants, as much as Graham Stuart Thomas . . . perhaps it was different.'

In the 1980s, Thomas was horticultural editor at Dent's, her publishers for *The Dry Garden* and *The Damp Garden*. While she had relied heavily on her notebooks for *The Damp Garden*, he suggested that her next book should be something different from the others, more personal, in the way of a diary. She should be inspired, he felt, by the Edwardian author Mrs Earle, who at the beginning of the twentieth century had had enormous success with her three volumes of *Pot-pourri from a Surrey Garden*. Something in that style would show Beth as a wife, mother and grandmother as well as a businesswoman, a woman with domestic responsibilities. He knew from Beth that she was having to cope not only with the garden and the nursery but more personally with the news that her daughter Mary's first child, Thomas, had been born with severe physical handicaps. He was to endure many operations in attempts to ease the fusion of his joints and give him some mobility. During these operations, Beth, most unusually, found herself unable to garden, anxiously waiting for news of how things had gone.

Although Beth had known Gertrude Jekyll's books since Cedric Morris had lent her his copies, she had not come across Mrs Earle's writings. By a lucky coincidence, on a trip to Germany to stay with Countess Helen von Stein-Zeppelin in Laufen, Beth found three copies of *Pot-pourri from a Surrey Garden* on her bookshelves, inherited by the Countess from her mother. While Beth knew that she and Mrs Earle came from very different worlds, as well as different centuries, she found the idea of a yearly record of her life and gardening appealing. *Beth Chatto's Notebook* became a long project, taking her over two years to complete. For many, it is a favourite for its personal notes.

Nearly twenty years older than Beth, Countess Helen von Stein-Zeppelin had trained as a landscape designer in the 1920s and learned about English gardeners such as Gertrude Jekyll from her plant-obsessed mother. In her twenties, she had inherited an estate in Laufen, south-west Germany, near the border with France and Switzerland, and begun to establish a nursery specialising particularly in her favourite plants, bearded irises. The Anglophile, bilingual Countess made regular trips to England. By the 1970s, her nursery was internationally famous for its selection of hardy plants, refined by her and her head gardener, Isbert Preussler. She had met Preussler while he was working in East Germany for plantsman Karl Foerster and soon offered him a job. Taking just what they could carry, he and

Countess von Stein-Zeppelin with her dog, a wolf spitz, at her nursery at Laufen.

his family escaped to the West and stayed with the Countess until he retired. Alan Bloom at Bressingham introduced Beth to the Countess in 1974, rightly guessing that this would become a friendship made in plant heaven. Beth's first visited the Countess in July 1974 during a holiday with Hans. They had driven across Europe in a Volkswagen camper van, through Belgium and France, arriving in the Black Forest forty-eight hours later. Beth was enchanted by the countryside. They camped high in the forest between two walls of cut logs and woke, as she noted in her diary, by a 'brook bubbling among rocks, ferns, foxgloves, wind in high beech and spruce'. They reached Laufen in the afternoon and were greeted warmly by the Countess. 'She was charming,' noted Beth, 'and invited us into her 300-year-old home to tea, asked me to write in her visitors' book.' Beth was impressed with the nursery and her lifetime friendship with the Countess and Isbert Preussler (who spoke no English) was cemented through that visit and a mutual love of plants.

BETH HAD FIRST GOT TO KNOW Alan Bloom in the late 1960s and had enormous respect for him. She was thrilled to receive a letter from him after several years of meeting, saying he would prefer she used his first name. Blooms of Bressingham was, she said, the one nursery she felt she could work in. The warmth was clearly reciprocal. Beth described arriving one day to be greeted by Alan 'already waiting for me with a barrow of empty boxes, labels, rubber bands, and polythene bags, and a spade to give me his treasures. His kindness completely revitalised me.'

Beth was always writing – lists, diaries, plant and travel notebooks – her natural talent had been nurtured at school and college. As part of her drive to promote the nursery when she opened it in 1967, she seized the opportunity to do a regular column in the local newspaper, the *Essex County Standard*, on anything from 'Beating the weeds with beautiful heathers' to 'Exotic leaves in frost and snow'. Beth wrote these articles to publicise the nursery but then she began to get

RIGHT ABOVE Beth, photographed by Hans, on holiday among the alpine wild flowers.
RIGHT BELOW Hans seated by the breakfast table complete with small cup of flowers.

requests from other sources. To begin with, she had no idea what to charge. Asked by Michael Hoog of Van Tubergen's nursery to write an article with some plans for their catalogue, she accepted without agreeing a fee. She later asked around for advice on what to charge, from Pamela Underwood and Alan Bloom. Mrs Underwood stunned Beth by saying the article was worth £250 and that top flower arranger Violet Stevenson would not have done it for less than £500. Beth was always keen to read books by interesting gardening writers. Margery Fish, whom she never met, was a favourite, as was Gertrude Jekyll. Her brother Seley gave her a copy of Christo's *Foliage Plants* when she was recovering from an operation on her thyroid in spring 1973. She devoured it, making notes in her diary as she read. 'I think I could write a book one day.'

Her first book, *The Dry Garden*, had come about as a direct result of Graham Rose's article on her in *The Sunday Times*. After the article appeared, her postbag was heavy with letters asking for advice on a plant for this place or that height. Throughout her life Beth made a point of sending handwritten replies to every correspondent, but these were taking up so much time that it seemed sensible to write a book which would answer many of these queries. That autumn, she was taken to lunch by two editors from Dent's who persuaded her to write 'a small book on "The Dry Garden"'. Once again, the problem was time – as well as the fact that she was more used to writing in the condensed form needed for articles.

Starting in January 1977, at the same time as she was planning for Chelsea, Beth turned to Christo's writing for inspiration. 'He writes so well, so very wittily.' She surprised herself by finding that she quickly adapted her style of writing, introducing examples to colour what she was saying. Her aim was to explain how to handle the soil. In areas of very low rainfall and light soil, the water drains away as through a colander. The challenge was to illustrate the sort of plants one could expect to thrive in such conditions. It took her just ten weeks to write the initial fifty thousand words. But it was to be a year before she held copies of her first book in her hand. 'What a thrill,' she wrote in her diary in April 1978. 'It really looks attractive and I would like the time to sit alone to skip through it . . . Instead I have battled all day with a diabolical NE wind, planting up the Holly Bed.'

The aesthetics of a book were always going to be important to Beth, but there were the usual budget constraints. She started drawing her own plans. 'It's strange',

Beth's first book, *The Dry Garden*, published in 1978, with its distinctive cover design by Margaret Davies.

she wrote, 'that I should be making plans for a book, when for years I said I can't plot a garden on paper.' It was decided that the book should have black and white photographs and line drawings. A word with Cedric Morris and artist John Nash (who had been to White Barn House several times with Cedric) produced the suggestion of Margaret Davies, who had taught botanical drawing and painting classes at Flatford Mill at Dedham Vale, just 'a crow's fly' from Beth's. They arranged to meet in October 1977 and got on immediately. Beth went armed with her ideas and quickly realised that Davies understood her desire for a portrayal of

June 1978 – 'The day I signed my book at Wivenhoe Book Shop – just after Chelsea!'

the plants in sketch form rather than a studied, realistic depiction. It was to be the start of a long partnership and friendship. *The Dry Garden* was an immediate hit. It was quickly brought out in paperback and reprinted four times in the early 1980s. It remains a gardening classic.

The natural progression was to write a companion book about the creation of what was called 'the damp garden'. It was here that Beth had dammed the water rising from the underground springs, creating the linked, tranquil ponds. She was keen to write not just about plants that need their feet in boggy soil around the edges of the ponds but also those that benefit from enough moisture to prevent them becoming just brown, raggedy stems as they would do in the gravel soil elsewhere in the garden – plants such as goldenrods, asters, echinaceas, the late autumn American daisies which come from the deep grass prairies with their rich black soil.

Again, Margaret Davies was commissioned to do the line drawings. As when they were working on *The Dry Garden*, Beth produced rough sketches which Davies would work up into a finished illustration. The front cover illustration was key. Beth gave Davies plants to take to her home in Buckinghamshire, where she worked on the delicate drawings. 'The dust cover is finished, I think,' she wrote to Beth in November 1980, 'richer in colour than when you saw it, and the Creeping Jenny adds a golden footnote to it. It is still growing in its pot and no doubt is looking forward to rejoining its companions at White Barn House.'

THE ENFORCED BREAK in the nursing home for her hysterectomy in March 1982 lasted ten days. Frustrated, Beth passed the time watching the daffodil buds opening from her hospital window. She received many good wishes but was particularly touched and astounded to receive a huge and beautifully arranged bouquet of gladioli, roses, lilies, double stocks, gerberas, iris and alstroemerias – all in shades of cream, gold, orange and red – from George and Olivia Harrison with a note saying, 'Chelsea is not the same without you! Hope you are feeling stronger & enjoying this lovely spring.' But the real thrill was to get back to White Barn House and be in her own garden. 'Walked immediately round bulb bed and Reservoir [Garden] with Andrew. Great improvement in 10 days – masses of daffodils now, *Muscari azureum*, chionodoxas, scillas, *Anemone blanda*, and *Pachyphragma macrophyllum* looking wonderful with so many yellow and bronze foliages.'

In May, Beth was back staying at Great Dixter. It was a cold spring and she went sensibly armed with plenty of warm clothes. She did not mind the spartan conditions of Dixter, the wooden floors, the whole atmosphere and smell of the house, the absence of 'interior decorating' – everything still much as it must have been when Christo was born, except for two new steel sinks, one in the kitchen, the other in the pantry. There were Calor gas heaters in the Yeoman's Hall which provided enough heat to write by and always a rug to wrap up in, which made Beth feel like being at home. In between garden visits and social dinners, Beth and Christo managed to find time to do some writing, an article for *Reader's Digest* on architectural plants by her, and a piece on the wild garden by him.

Beth marvelled at how quickly and freely Christo wrote. He claimed it was just practice but Beth admired his direct, flowing style, interspersed with occasional caustic wit. 'Living with him,' she later wrote, 'I can see how his everyday speech and events are drawn naturally into his articles.' Also she envied his self-taught touch-typing skills and regretted not persisting with those herself. Later that afternoon after lunch, Christo suggested they go back to the Parlour (his study) where there was a smoky log fire, as it was the only warm place to have their coffee. It was so cold that day that the house guides, waiting to show visitors around, were seen shivering in the main hall. As they approached the Parlour, they were met by a party of visitors just entering the room. Beth held back, clutching Christo's elderly dachshund, Crocus, but Christo swept in, so she had no choice but to follow. Whereupon the pair sat, sipping coffee, like part of the exhibit, while the old samplers, china, chairs and bookcases surrounding them were explained to the visitors.

The next day, Beth reluctantly let Christo read her finished article. She had found it hard to concentrate long enough to get past what she called 'the sticky patches'. She read some of his articles and was able to spot some tiny spelling errors. But the thrill for her was to sit in his study and write with him, trying not to be distracted by the books that took her fancy: a large catalogue of a recent Lutyens exhibition, a book by Christo's father on yew and box hedge topiary, showing the garden at Dixter as it began, and a book on perennials written by Christo fifteen years earlier. This last was, Beth thought, full of facts and good sense, plus several

Beth on one of her regular visits to Great Dixter, with Christo and Tulipa.

laugh-out-loud passages. But Christo said it was not selling well, too much small print perhaps. 'People are too lazy', he said, 'to dig in, either into print or the soil.'

Beth's second book, *The Damp Garden*, was launched in autumn 1982. In his review of the book in his *Country Life* column, Christo said of Beth that 'had she known at the time of writing *The Dry Garden* that there would be a damp-gardening epic to follow, she would have planned the former differently.' He never held back with honest criticism even of one of his closest friends. He was no fan of the pulverised bark mulches she advocated. 'Any photograph of Beth Chatto's garden is instantly identifiable. There are the bark mulches. They are not ugly. . . but they are ever so slightly disturbing.' What he loved about the book were 'its value judgements and its plant descriptions'. As always with Beth's writing, Christo knew that 'Beth can hold a sprig in her hand and immediately put into words . . . the essence of that flower, leaf or plant.'

Beth's parents, Bessie and William Little, in the garden of their home in Wivenhoe, October 1972.

While she may have been buoyed by good reviews for *The Damp Garden*, the many empty pages in her diary that year are a testament to the emotions she was dealing with in her personal life. The health of her mother, now ninety-six, was failing. She and her husband had been moved into a care home but placed in separate rooms. In November, Bessie Little was admitted to hospital. Beth's father, William, at eighty-four younger than Bessie by some twelve years, was distraught. For sixty years, it had been a close and happy marriage, though William had suffered bouts of depression later in life. That spring, Beth had made one of her regular visits to see them at their home in Wivenhoe. 'I picked and made two posies of flowers for Mother. It was wonderful to see her looking amazingly alert, like a tiny bird among her shawls and rugs. She was elated by the flowers insisting on guessing the names of most of them . . . Father, sadly, could not be roused by any of them.'

Shortly after it became clear that her mother would not recover, Beth was told that her father had fallen from an upstairs window at the home and was in intensive care. He died later that day. Three weeks later, Bessie Little also died. In a letter to Christo shortly before her mother's death, Beth wrote of her courage though so old and frail. 'She makes us smile as she busily stage manages the funerals to come (exasperated that there will have to be two!) discussing the hymns and music (nothing mournful).' But their deaths, and in particular the fact that her father had committed suicide, hit Beth hard. 'I sometimes fear', she wrote to Christo, 'I shall never again be able to write something so involved as a book.'

The blank pages in Beth's diary that year are also testament to a change in her relationship with Hans Pluygers. Perhaps put out by Beth's increasing fame and unavailability, he had started seeing another woman. Beth was devastated, although she knew she was in no position to demand fidelity. It had always been clear that she would never leave Andrew, particularly as his health was deteriorating. But she also could not bring herself to break free from the Dutch farmer whose personality and intellect could not have been more opposite to her husband's. For more than twenty years, the arrangement, in its own strange way, had worked – most of the time. And there was always one thing this current crisis could not change. Their lives were inextricably linked since their lands shared adjoining boundaries. Beth's regular evening visits to Park Farm stopped, but once the dust settled, a truce was called. Hans now became an occasional visitor to White Barn House, seeking Beth out but still steering clear of Andrew, as a new era of a sometimes strained friendship developed.

BACK AT CHELSEA IN 1983, Beth's exhibit drew more praise and yet another Gold Medal. Rosemary Verey, reporting on the show for *The Garden*, commented that hardy perennial stands, particularly Beth's, 'seemed to attract the densest crowds, all earnestly taking notes.' For many, it is that image of Beth's stand that stayed with them: visitors five or six deep all straining to see the names of Beth's new finds, scribbling them down or ticking on the order forms being handed out by Beth's staff. Beth herself was, as always, mobbed with a constant queue of eager questioners including a young Carol Klein, now a well-known BBC television

One of Beth's Gold Medal winning stands at Chelsea: 'I did not aim for a showy display of unseasonal plants in flower but tended to concentrate on the effect of contrasting shapes, sizes and designs of leaves.'

gardening presenter, who never forgot Beth's patience in answering every query. Verey, very much part of the zeitgeist, wrote that Beth's colour combinations of yellows, white, greys and mauves moving on into plantings of brighter colours were 'perfect'. 'It is right that Beth's artistry should have such influence amongst present-day gardeners; we are able and willing to devote as much time and attention to the colours of our borders as we are to the curtains in our house.'

It had been a slow start to the show that year because of the rain. Beth also noticed a reluctance by customers to order on the day. She felt that she signed more autographs than she took orders, but it soon became clear that this was because customers were enthralled by her detailed catalogues and wanted to study them at home first. With bulging order books that came later from the Chelsea shows, propagating became a seven-day occupation for Beth (Sunday was the one day of the week she could work uninterrupted potting up hundreds of seedlings). During the week, her small team of girls was stretched as well. Emotions often ran high as personalities clashed. According to her diary, Beth spent a great deal of time smoothing ruffled feathers or listening to stories of broken relationships with sympathy and cups of tea, 'silently relating it to my own situation'.

The problem of needing an estate manager had been solved. In spring 1978, she had employed Keith Page to help Harry Lambert with the heavy work around the grounds. Thirty-four years earlier, Page had been the schoolboy who challenged Beth about the right way to do things in the garden and was then instructed by Beth to do things her way. After school, he had joined the RAF and done engineering. He had then run a small market garden as his father had done. He was to become invaluable to Beth as he took on the role of estate manager, maintaining everything from tools to polytunnels until his retirement.

Beth was always on the look-out for enthusiastic help. In the early 1980s, she had noticed a small group of young men visiting the gardens, students who came from the Blooms of Bressingham nursery regularly to study the nursery and its plants. One was soon taken on as a propagator and stayed for a while. But it was another one who Beth felt would suit the nursery 'family' best, a young man called David Ward. Although from Norfolk and therefore an East Anglian by birth, Ward had gone off to work for Avon Bulbs in Bath. Beth wrote and asked him if he would come and join the team. Held back by loyalty, it took Ward a year before he relented and returned to his roots, starting with Beth on her sixtieth birthday in June 1983. It was a good decision for both of them. David Ward remains at the nursery to this day. Writing in her diary on 27 June 1991, Beth noted it was 'David Ward's eighth anniversary of coming to work with us. I think it was the best birthday present I could have had – and still think so, every day.'

TRAVEL NOTES 2

Beth's first trip to the US (1983)

IN OCTOBER 1983, Beth left for America for her first major foreign lecture tour. It was to launch her in the New World, a world she was not used to. Her nearly three-week trip covered four states – New York, Massachusetts, Pennsylvania and North Carolina. From the moment she boarded the Heathrow to Newark People's Express Boeing 747, to her return nineteen days later, she tried to record every garden visit, every notable plant, every lecture event, every host's home, every meal eaten, even every outfit worn, filling several of her favourite red Silvine notebooks. She was rising early but going to bed later than usual, with only one phone call during the entire trip to keep her in touch with the garden and nursery and to check on Andrew's health.

The trip was full of memorable moments, from having a bath in brown peat-stained water in wealthy Frank Cabot's New York estate, to seeing Fred McGourty's* cold frames sunk five feet deep to protect plants against the winter weather and flying to Philadelphia over the forested Appalachian mountains vivid with autumn colour – scarlet and lemon maples interspersed with pine.

Beth's hectic programme included four lectures to large audiences at the New York Horticultural Society, the Arnold Arboretum in Boston, the Pennsylvania Horticultural Society in Philadelpha and the North Carolina Botanical Society in Raleigh. She also met some of the grand figures of the US horticultural world such as Lincoln Foster,* father of the American Alpine Society and J. C. Raulston* of his eponymous arboretum at North Carolina State University. In between, much of the time, she was travelling, visiting gardens and sightseeing, even including a trip to Trump Tower. As with all trips of this type, it was exhausting but for Beth also magical, as these diary selections show.

It is almost 10 a.m. I am sitting in Seat no. 40G on Flight PE001 Newark, New Jersey. My seat belt fastened, the safety instructions have been read and we are ready for take-off. The Captain has just said 'sit back enjoy yourselves – it's a lovely day for flying' – and indeed it is – a beautiful calm sunny autumn morning. The seat next to me is empty, but beyond it is a pleasant American woman just returning from her first trip to England. She loved it – it was so GREEN. 'Is it like that all year round?' she asked.

57 degrees in NY. Cloudy I think. I wonder how long disembarkation will take? How soon before I shall see Larry Pardue,* Director of the NYHS [New York Horticultural Society]. I shall carry a copy of *The Dry Garden*. The pinger goes again – people are being asked to stop smoking. We are about to land.

When eventually I arrived at customs I saw a girl ahead of me have her luggage opened and immediately before another was asked if she had any plants or seeds! I thought of my seed packets but when asked why I was here replied lecturing on garden design! The official beamed and said his wife would be interested and forgot to ask any awkward questions. Arrived at the office at 8.30 a.m. but locked (Larry Pardue is there from 7.30 a.m. but tucked away upstairs).

Soon 9 a.m. arrived at the office, gradually people came, secretaries – young men and women, some coloured, all very charming and welcoming to me. When the coffee was brewed, we sat and talked. Mr Charles Webster,* an elderly man, reminded me of John Codrington* but less bushy, obviously an important figure – an ex-director perhaps [he was the chairman].

12 p.m. Set off with Larry Pardue for a tour of the city gardens (not botanic). I was much impressed by the community gardens. The NYHS has about 500 of them. The sites were originally miserable rat-infested dumps. Now they give opportunity of expression, creativity, colour and food. The one I saw, Westside Community Garden, had been designed by the people who care for it – the beds laid out in the shape of a gigantic sunflower to be seen from the high-rise flats around. These flats were occupied by very poor people but across the road quite a different community – people with quite high salaries – yet all worked together on the gardens. A great leveller, said Larry. Mostly the plants were annual

bedding, cosmos, alyssum, dahlias, tagetes – but I also saw *Salvia officinalis*, *Lavandula angustifolia*, *Stachys lanata*, *Salvia turkestanica*. The vegetable plot, though shabby at the time of year, was neatly laid out with paths and long narrow borders. It was a most moving sight I found. The NYHS does not run or organise them but is just there to intervene and help when help is needed.

The garden, Vale Hill Gardens, was in some way connected with the family of Frank Cabot's* wife and is one of the most interesting gardens of New York. It is small but full of interest – with a fine view across the Hudson to wooded slopes all shades of autumn colour. I was introduced to the great Tom Everett* who is the [grand] old man of American Horticulture. Originally from Yorkshire I think – he has written many books including an encyclopaedia of gardening which runs into 10 vols. He was complimentary about *The Damp Garden*. I felt very honoured.

Dark and drizzly we . . . drove northwards about 70 miles to Francis Cabot's place [Stonecrop]. [Next morning], I found Frank (as he likes to be called) sitting at the round table in the warm (thank goodness) kitchen, sorting the seeds I had brought. Borrowing boots, we set off into the garden. A bad time of year to see the garden, but I found it strange. Herbaceous plants not well displayed . . . Interested by Belgian espaliered fence, perfectly trained into diamond shapes.

It is obviously difficult to grow many plants including alpines in the climate here, so Frank has many forms of shelter. First I was shown the half-sunk greenhouses where all kinds of tiny things were growing. In flower I saw *Narcissus elegans*, *N. serotinus* and *N. viridiflora*, and cyclamen including *C. rohlfsianum*. Elsewhere were other complicated and expensive machine-controlled buildings – even to reduce summer temp to 30 degrees when daytime could be 90 degrees for his *Kabschia* saxifrages.

Frank is obviously a collector of many things – plants, fine furniture, paintings, china, etc. He has started a large rock garden – with expensive plans to make a great waterfall – elsewhere a tranquil little stream meanders through an alpine like meadow where the forest has been cut away to expose it. I had the feeling that less money and more experience and personal involvement – plus a lifetime – was really needed to make the garden satisfying. (I was very pleased to meet one of his girls, Betty, who works in the garden and nursery – he sells a few plants to some visitors – and to see that she had potted up other plants which F had collected from our nursery only the week before.)

"Unusual Plants," Colchester

THE HORTICULTURAL SOCIETY OF NEW YORK

LECTURER BETH CHATTO

October 19 – November 3, 1983

Changed into my Laura Ashley dress. Larry Pardue arrived to drive me back to New York.

From 4 p.m. to 6 p.m., I sat with the audiences through the annual general meeting. My heart beating furiously throughout, it was an ordeal. I sat beside Connie Gibson, a charming lady aged 80 or so who was to be my hostess for the evening. I took to her immediately. There was an endlessly long interval – but at last I could begin, desperately keyed up. But I had been filled with emotion at presenting the prizes to the winners of the community garden projects – the recipients had behaved so well, such simple dignity and simple speeches of thanks for the help given by the NYHS – that it gave me an opening speech – of appreciation of all involved. That off my chest, I managed to take myself and the audience back with me to Elmstead Market – and the rest went like a rocket. I enjoyed myself, words flowed, the pictures looked good and at the end I had a prolonged and hearty applause. Charles Webster – who had previously treated me with the polite reserve one has for a stranger – gave the vote of thanks then came forward took my hands and kissed me! More applause! People lined up to shake my hand and say personal thank yous. It was overwhelming.

Thursday 20 October 1983

After breakfast I walked round Gramercy Park. Along one side is the house where Lanning Roper* was born. I went and looked, feeling sad I could not tell him I had stood there. I passed by the National Arts Club and just popped in again to see the beautiful arrangements of *Celastrus scandens* – long twisting ropes of orange and cream berries simply arranged in fine Japanese vases on huge mantelpieces against huge mirrors.

Next we visited the Trump Tower. This is the latest fantastic attraction for New Yorkers as well as visitors. Designed and built by a young man, it epitomises in theatrical fashion the western obsession with selling and spending. We were the actors on a fantastic stage – moving staircases with glittering brass handrails watching ourselves in walled mirrors as we sailed ever upwards around the central well of this vast tower. All the walls are marble, in warm shades of pinky-orange, which the brass picks up – contrast is made with masses of handsome foliage plants strategically placed. At one end, the entire wall, from top to bottom, consists of a stepped waterfall – each step lit so that the cascading water catches the light before its next sheer drop. At its base, on the ground floor, is a restaurant – we could look down on the ever-diminishing little tables and parasols and hear faint music coming from nowhere. On every floor were boutiques, shops, of jewels, furniture, fine clothes, leatherwork. I was relieved not to have to make any choice in such a place. It was just enough of an experience to be there.

Friday 21 October 1983

I changed into my only dress to go to luncheon (not lunch!) with Mrs Adele Lovett,* a great character and wife of some high member of government – perhaps secretary of state [Secretary of Defense]. She, aged 83, was memorable. She was dressed entirely in pink – hat, suit, blouse – all shades of pink and cyclamen with pearls and beautifully manicured nails (Barbara Cartland you might think but definitely not). Yet gardening is obviously her passion (or one of them!). We were met on the porch and immediately taken on tour. I was struck by vast carpets of pachysandra under the trees each piece planted by her. Walking in the box-edged herbaceous borders she talked of her friendship with Constance Spry – showed me her old-fashioned roses. Indoors, a maid in black with lace edged cuffs and pinny waited at table. Finger bowls at each place were scented with *Lippia citriodora* and sweet rose geranium leaves. (I have just been reading exactly that in Mrs Earle's *Pot-pourri from a Surrey Garden* written almost 90 years ago.)

The bathroom, blue and white – white towels with blue monograms, flowered china toilet bowl. I could not resist using the perfume spray before I left!

Beth flew to Boston for a visit and lecture at the Arnold Arboretum.

I am beginning to like flying, seeing cities, fields, woods and roadways spread below me – all the watery inlets, the sandy edges of the coastline against the inky blue sea. On landing, Mary Ashton,* wife of the director of the Arnold Arboretum, was standing with my name on a card.

Saturday 22 October 1983

Woke to sharp white frost everywhere. We were scheduled for, unfortunately, a short sharp drive round the arboretum because it had been arranged that I talk on the telephone to some radio station at 1 p.m. But nevertheless, it was a thrill for me to see where E. H. Wilson* had walked and worked, having read his *Plant Hunting* last winter.

My general impression after such a brief and hurried drive round, two hours maybe, is not a fair thing to attempt – I remember being taken to a height overlooking a fine view where Harvard University and the building I talked in last night were pointed out to me in the distance among all the buildings of Boston – well, of course, it is nothing like Kew, in that it is more natural – entirely trees and shrubs in grass, hardly any dug beds, which give a more artificial look. Although there are many exotics introduced from the east, the effects are uncontrived, blending with the natural forest of the area. But like many such places it has depressed periods and young Mr Koller* is bent on making improvements, replacing losses caused through neglect, and doubtless introducing new subjects. I am not sufficiently knowledgeable about trees and shrubs to appreciate all that could be seen there.

Before dusk, we drove around the neighbourhood through entirely woodland or forested roads. We passed a Hutterite community originating from Germany, all the inhabitants are completely self-contained and maintained, still wearing the costume of the 19th century.

By the roadside we saw Christmas fern, *Polystichum munitum*, *Tiarella cordifolia*, *Celastrus orbiculatus* in good fruit gone wild, and *Berberis vulgaris* and *B. thunbergii* which have become a nuisance. Trilliums grow in the woods and cypripediums. Once or twice, Fred [McGourty] said well done because I recognised some withered seed head! I felt I was being accepted. He is possibly rather shy. He appears reserved, suspicious perhaps. But obviously loves to talk plants and is very well informed, more widely read I would think than I am!

We then retired to the study and Fred showed me his talk at the Arboretum. There were some good slides, but not, to my mind, artistic, or particularly connected. The first slide was the best, a view of the garden in the spring with a long view to the house backed by trees with a glorious foreground of *Cardamine pratensis* in the grass. Finally we ate ice cream with maple syrup and walnuts by the fire with three cats to make up the company. A very happy and interesting day.

Sunday 23 October 1983

I started this trip a week ago today. Thankful my bedroom is warm. Woke 6 a.m., got up and made a cup of herb tea, back to bed to write up notes and bath. Fred was cooking breakfast. I asked what I could do. He said sit and talk to me (we did that for the rest of the day!). He cooked crisp sweet bacon, drained on kitchen paper, with blueberry pancakes served with maple syrup mmm! It was delicious

– reminded me a little of the large thin pancake I have had in Holland, on a chilly autumn day, spread with syrup and broken pieces of crisp bacon. There had been 12 degrees of frost. Then we started a day of visiting.

First we called in to John Saladino,* a top New York interior designer, at his country house for weekends and summer. Grandiose in style, Palladio perhaps? All in shades of grey and mulberry with high archways and pillars – vast plants – trees in fact in tubs. Enormous tubs of agapanthus in the dining room. We drank coffee by a welcome fire after having viewed the garden. The place was derelict when he began, only a few years ago. The best he has done is to thin some trees and shrubs around the house – not too many because they are kalmias, *Euonymus alatus*, wild azaleas as well as superb red maple, with brilliant sugar maple dominating mixed woodland which fills the valley as it drops below the house and rises again in the distance.

Being Italian, he pines for grey plants. Both attempts a disaster I thought. One a messy *Stachys lanata* and other grey-leaved plants under trees being obliterated with huge wet leaves – the other a long limestone pavement starting and going nowhere, with a few creeping things in the cracks – what a lot of expense so ill used. Where was the feeling for harmonious design – all left behind in the house? Poor man – however he was so full of enthusiasm, though I doubt able to afford such mistakes – if he cares enough and goes to see better gardens he will succeed one day – he has a site which could be ravishing.

Next was a visit to Ted Childs.* As we drove, Fred told me he was an engineer by profession, but also owned thousands of acres of forest hills we were driving through. He is the wealthiest man in the district, makes maple syrup with his foresters on the estate and has lived and travelled much abroad in his youth.

He reminded me a little of a fox, our old fruit growing mentor and guide. Tall, well-built, kindly, humorous-looking, aged almost 80 years. Wearing comfortable almost shabby clothes, he walked out in a slight drizzle to greet us. Behind his fine green shuttered, white boarded house you walk to the edge of the cliff, where you look down on his collection of alpines – dwarf shrubs and conifers. It has apparently begun from a farm yard with wooden barns, piggery and a sloping walk down for the cows to the valley pasture below. Now the cliff garden passes into regenerating woodland rising gently on the opposite slopes. This garden is a new project for him, only about 15 years old. But I thought it the most sensitive piece of planting and design I had seen so far. Like me he loves foliage – to the extent of cutting off the seed heads of *Dryas octopetala*! I teased him about that – I think they are so beautiful. Every ledge was covered with chippings, the rocks and steps all creating comfortable niches and crevices for plants which were well grown and comfortable – with not a weed anywhere.

The site was open to the sky above but pines and other trees created a backcloth, as did an old barn-like building which had a kind of tower at the end. The whole smothered in plants, including *Hydrangea petiolaris* and *Schizophragma hydrangeoides*. As you look up the slope this made a splendid focal point high above. The garden had a natural harmonious look, even though much hard work re-siting stones, pathways and terrace beds had been done. A fine stone-filled gully had been built full of rounded boulders and large stones. Dry now, this gulch takes melting snow water from the land above (100 inches of snow yearly). Not bothered by his age, at the bottom of his cliff garden, was a new area planted with young trees and shrubs, on the edge of this meadowland and woods. On a shady bank was a collection of native shade and cover plants. Tiarella, *Cornus canadensis*, trilliums, etc., and another little patch of native bog plants. It was very impressive because it was so sensitively designed and planted and yet as a young garden, no feeling at all of an old man. It drizzled with rain the whole time, but we had umbrellas and scarcely noticed. He was such a nice old man, so loved his plants. Each one was inconspicuously labelled with metal labels. I should like to meet him and see his garden about once a month.

Tuesday 25 October 1983

Through Torrington and New Britain we went until we came to Sunny Border Nurseries, Kensington, Connecticut, the home of Pierre Bennerup.* It was more than good of him and his wife to see me as his mother had died only about a week previously. Beautiful flower arrangements filled the house which was some way from the nursery, a tall gracious Victorian house, although still white-boarded and black-shuttered. We were invited in immediately, into a room warmed by a large, black and handsome wood stove, smelling of good coffee, with two huge plates of pastry cook goodies.

It was good to see Pierre because I first met him on my own nursery not so many weeks after my hysterectomy Arrived at 9 a.m. one fine morning around June, looked all round then sought me out to talk hostas. We went down to look at my special ones – he was full of enthusiasm. Then he went back to Bressingham where he was staying. At 2 o'clock the phone rang. It was Adrian Bloom.* Could he come with Pierre again to see the garden? I was delighted and wickedly gleeful – sure that Pierre had gone back bubbling about my plants to find that Adrian, one and a quarter hours away, had never been here. To be fair to Adrian, he runs a vast empire of a plant business so simply had not found the time to come although we have met at Chelsea and at Bressingham.

Pierre runs a wholesale nursery for herbaceous plants. He employs 16 people now, in late autumn, more in the summer, has 12 acres of land with 80% of stock standing in pots, the rest stock beds. Being wholesale, it was not charming to look at (not so tidy and efficient looking as Blooms either). The site was not tidied at all, plants standing in soil, no gravel, so pots get mud splashed. No fixed irrigation. But I was interested to see the small micropropagation house with two young girls working in it. They had hostas in test tubes, not looking too bright. What is the next step, weaning and growing in house? I didn't see that.

We had whisked round Pierre's garden, saw lots of hostas. He gave me a few – hope I can keep them alive – and also gave me several plants of a silver yucca that I saw in his garden – most handsome which would be really worth introducing called *Yucca glauca* (Graham Stuart Thomas lists it in *Perennial Garden Plants* but says it is a rare species seldom seen in this country, probably needs almost desert conditions in our warmest counties to make it flower) – we should be able to manage that. I would be content if I could just grow a fine plant without flower – they looked outstanding in Pierre's garden where it must be infinitely colder in winter, although much hotter in summer.

Tuesday 25 October 1983 (12 p.m.)

Fred has just deposited me at the airport outside Hartford. I feel and look like an immigrant with my bags of slides, plants – with hosta seed pods sticking out of the top, my suitcase and handbag! A smart blonde woman, clad head to foot in navy, including gloves, makes my lovat green tweeds feel very countrified. Neither do I see anyone else with mud on their shoes.

Beth was met in Philadelphia by her hostess, Kathie Buchanan,* from the Pennsylvania Horticultural Society.

When I had first seen Kathie's car with an artificial chrysanthemum fastened to the radio mast I had, superciliously I confess, thought oh dear! I do not like artificial flowers anywhere and would not think of having one as a pennant on my car. But as we walked out into the gleaming mass of cars I suddenly saw the point, relieved to find the car so easily and laughed at my silly attitude. As the seasons change, so does the flower, a daffodil in Spring, a rose in Summer.

She is an amusing lively lady. We talk well together on all kinds of things women encounter. Their children, husbands, travel, books, house decorating – and the state of the world, last perhaps because we can do least about it. I think it is good for women of all ages to be able to talk with their contemporaries, to discover that we are mostly experiencing much the same kind of problems and pleasures, to realise that we are not alone. To know you are sharing a common experience helps to balance your outlook, so that you may laugh rather than frown.

[The following day,] we were driving to Bryn Mawr to have lunch in the dining room of the well-known college there which was originally for women only but is now mixed. Among the famous names which have studied there was Katharine Hepburn. As we walked across the grass back to the car I had to go and investigate the huge ginkgo tree which I had noticed from the dining room. It was a female tree and the ground below was littered with fruit, like small golden-brown round plums. They have a filthy smell, but I tasted them and found it not at all unpleasant, just sweet and juicy. I collected a handful, found a plastic bag and put the evil smelling parcel into the boot of Kathie's car. She was very tolerant.

That evening, Beth gave a lecture to the Pennsylvania Horticultural Society in Philadelphia, the second-oldest horticultural society in the US.

I don't know the first – was it New York? [No, it was the Massachusetts HS.] I was taken to the library where I was asked to sign copies of my books. I find it very strange to be treated like a VIP, yet aware that I was taking part in a special occasion and as I happen, for a short while, to be in a central role, I must play the part. But acting doesn't come into it when you feel you are among friends. Although newly met we all recognised each other bonded by a common love of plants and natural beauty. I was reminded of my dear friend Cedric Morris's remark, made several times in the last years of his life, that gardening is the only civilised thing left to do.

I found myself sitting with Jane Pepper,* director of the PHS, and a very interesting woman, Joanna Read. She is obviously a most enthusiastic plantswoman, someone with whom you know you could spend days talking about plants and then not dry up. Somehow she reminded me of the kind of American woman Joyce Grenfell would have portrayed so sensitively, rocking on her porch, talking, reminiscing in a gentle attractive accent.

Concerning the American accent, especially as we so often hear it harsh and grating, mangling the English language on film or television screen, I was relieved to find myself enjoying without exception the far less abrasive tones of the people of New England, at the same time appreciated the slight variations in accent that I could detect from New York to the Berkshires and from Philadelphia to North Carolina. Regional accents greatly enrich the language. They are colourful, emphasise individuality and often underline regional characteristics. Why do we like some accents and dislike others? For myself, I dislike speech which may be called slipshod, with consonants swallowed or omitted altogether and vowels distorted out of recognition. When languages become coarse and neglected there is not pleasure in listening to it, but a good voice has all the sensual delight of a good and varied meal, each syllable contributing to a feast of sound.

Jane made a mercifully brief introduction and after all the long wait, the mounting tension, I could begin.

It would be quite useless for me to have prepared any introductory speech on this occasion. After such strenuous entertaining, any well-considered phrases would be entirely wiped from my mind. However I began, it had to be spontaneous. Although incapable of telling jokes, something seemed to amuse them so, feeling their support, I tried to take my audience with me and journey back across the Atlantic to my East Anglian garden. One has to remember that the map of Great Britain is as vague to much of one's audience as the United States was to me before I began this trip. But fortunately Andrew had a good stock of *National Geographic* magazine maps, so I spent much of my time before I set forth on these adventures studying the regions in which I would be staying. Americans seem to know London, the West Country and Scotland. Anything east of London is as unimaginable as the Dark Ages, which is a pity because East Anglia was practically the only civilised part of the country in the Middle Ages with towns and villages developing through the flourishing wool trade. It was also an area within reach of London where wealthy statesmen and noblemen kept fine estates and helped to build beautiful churches. To this day, the centres of many East Anglian towns and villages show the gradual evolvement of domestic architecture throughout the centuries. The last 100 years have not, on the whole, added improvements but that is the case worldwide.

Although I felt they might be the most critically aware audience that I had, I also felt encouraged to linger with the last slide shown, the long applause which followed was like wine, as were all the kind comments and personal contacts. I was pleased to speak with a number of young people, among them a young man from the famous Winterthur Garden [near Philadelphia] – all of whom said they were inspired. How rewarding for me not just the acclaim, but to know that the long years, 40 years in fact, of teaching myself how to plant and group, of struggling with awkward soils and erratic climates, could be a help and encouragement to others. This is what makes gardening so satisfying, it is a caring sharing occupation. It is, I agree, a selfish occupation as well since one needs to spend much time alone to practise the art and craft. But it is a common language and bond. Wherever you go, if you recognise your plants in someone else's garden, you know that person would give you a welcome.

I would like to say that my slides were quite beautiful, some of them breathtakingly so. This is not so boastful as it sounds because I did not take any of them myself. As my family and friends know I am the most hopeless person with any kind of mechanical gadget – I cannot even type. Sometimes I wish I would make time to hold a camera steadily because there is not a day in the year when I could not find a dozen desirable views or plant portraits. But I usually fail to keep my garden diary up to date, the writing of articles or the occasional book is consistently squeezed between running the nursery, redesigning and replanting the garden, or being happily involved with preparation and people. So I am greatly indebted to both professional and amateur photographers who over many years have made for me fine slides which record the development of both garden and nursery.

Today we drove to Longwood Gardens [forty miles west of Philadelphia]. We were met by a handsome young woman, Landon Scarlett,* the curator, who designs and organises the great display gardens which are such a feature here. Not all was formality, we drove through areas which made me think of Sheffield Park [East Sussex] – areas of fine trees and shrubs in natural planting. *Pachysandra procumbens* was used as ground cover – I first saw it in Frank Cabot's garden – he gave me a piece. I hope I can keep it alive until I get home. We passed by *Magnolia acuminata*, which Landon said was the only tree (or plant) which Harold Hillier noticed. Being a native of the eastern US, I expect he was as thrilled to see it as I have been seeing these plants which have thrilled me in the wild – we were shown the Meadow Garden – about 20 acres. Not a good time to see it – but good to see that they are obviously involved in conservation and preservation of native species.

I don't think we have 20 acres of natural grassland in Great Britain where wild plants are being preserved or have we?

Finally we came to the inevitable shop, where I bought a book on Longwood Gardens for me and little painted glass pictures for my daughters. I was asked to sign my books, and at 4 p.m., after a most enjoyable day, said goodbye to Landon Scarlett who had given us such a private and privileged view, including the library and archives, potting sheds, growing on areas – even the labelling centre where we met a man producing labels photostatically!

Winterthur was 6 miles too far away – we both felt like home, so we made a pleasant drive in time for tea. I remembered my smelly parcel of seeds so I cleaned them in a bucket of water.

The whirlwind tour continued with a flight to North Carolina for more talks and a meeting with the NC Wild Flower Preservation Society at Ahoskie. But for Beth, the weekend highlight was a canoeing trip organised by her host, Ollie Adams,* to the Merchants Millpond State Park.

Saturday 5 November 1983

Together we spent a fantastically beautiful two days, canoeing among the taxodium and tupelo. Saw eupatorium, phytolacca – swamp plants we grow at home. Ollie thought we needed to be with someone experienced, but [Ollie's friend] Tom thought we could cope and so we did! Ollie sat in [the] rear, and wasn't quite sure how to turn left or right but eventually we found a rhythm, and it was the most thrilling, beautiful adventure.

There are hundreds of acres of water, black shallow water, which is forested with these swamp cypresses growing in it, and the tupelo trees, and plants that we do grow here in the garden. And there were logs floating and sitting on these were little black tortoises [turtles] with yellow feet, lifting them up to sort of warm them. As you canoed through the still black water your faces were brushed with the Spanish moss, which you sometimes see now sold in florists – long streamers of moss. Oh, it was magical really!

Beth with Ollie Adams and friends canoeing on Merchants Millpond, Datesville, North Carolina.

ENDNOTE

In 1984, Ollie Adams described that 'magical' day for members of the North Carolina Wild Flower Preservation Society in their spring newsletter:

> In September of 1983, I was on a committee exploring the possibility of having Beth Chatto, the famous English gardener, writer and nurserywoman who has won eight gold medals at the Chelsea Flower Show for her displays, present a lecture here in Raleigh during her first visit to the US. Even though time was short, we decided to try to get her for a date and I was delegated to make the call and extend the invitation.
>
> My name meant nothing to her but when I said hesitantly that we might be able to visit a cypress swamp in the eastern part of the state during her stay with me, I caught her interest. 'Do you mean *Taxodium*?' she inquired, and I, always a bit nervous when the telephone lines stretch across the ocean, said 'yes', relieved that she had known the botanical name that was eluding me at that moment!
>
> I was relieved again on Saturday 5 November, when we met Julie Moore in the woods at Merchants Mill Pond where we had arrived too late to participate in the morning's planned activities. Beth Chatto uses only botanical names when she discusses plants and my own knowledge was running a bit thin, for she had asked about every roadside weed and grass on the trip from Raleigh to Gates County!
>
> She adored our afternoon canoeing on the pond (first time for either of us to 'paddle our own canoe'). The day was near perfect and so warm that we took off our jackets and basked in the sun just as the yellow-bellied sliders [turtles] did on the half-submerged logs. They were so sun-baked and relaxed that they let us come inches from them before they slid off into the water. The combination of cypress hung with silver Spanish moss, yellow-leaved tupelos against the blue sky heavy with blue-purple fruit like giant olives, red-hipped swamp roses and aronia, and olive-green wax myrtle was a planting plan that couldn't have been improved upon. Beth took away with her a picture of a truly American garden, the kind that can't be seen in England.

8
Golden years

When Beth won her seventh Gold Medal at Chelsea in 1984, Arthur Hellyer, doyen of the horticultural world and by then in his eighties, writing in the RHS journal, noted with some sadness: 'Among the losses one must count the fabulous rock gardens that once occupied the whole of the Embankment side . . . no one ordered such gardens any more, they became merely prestige show pieces and they were gradually superseded by gardens of a more practical nature.' In the seven years since Beth had started exhibiting at Chelsea, fashions for plants had changed and Beth was, in many ways, a contributor to this change.

The year had started quietly, as Beth had the gardens and particularly the nursery to deal with. There were always staff dramas as the pressure grew with the ever-increasing order book. Alan Bloom warned Beth about letting the business get too big, to the point that it ran her rather than the other way round. She was concerned about that herself, but not sure how to control it.

Distraction came in March 1984 in the welcome form of one of her regular visits to Great Dixter. Beth was due to give a lecture at Wye College of Agriculture in Kent. With some large snowdrops and *Narcissus pallidus* tucked into her trug basket as gifts for Christo, she drove down to East Sussex, a route she knew well by now. The next day, she and Christo drove together to Wye, Beth wearing what she admitted was a rather strange outfit of a green tweed skirt and shirt and red waistcoat and stockings. She told Christo she felt nervous speaking in front of him, a notion he pooh-poohed, saying there was no reason – he wasn't the Queen. Beth insisted that he was far more nervous-making than the Queen. Later that evening, Christo produced a notebook in which he had made notes about Beth's lecture. He quoted her as saying, 'I don't stand still long enough to take photographs,' 'Aristocrats are the plants you cannot grow well, e.g., primulas,' and, to great amusement, 'Grit has

The gardens and nursery in 1984, much expanded but still surrounded by lines of apple trees.

made my life.' To Beth's relief, he said that she stood well ('straight and still – not rocking about') and that her delivery was perfect. 'So! I think I was approved. I'm glad,' she noted later that evening after a dinner hosted by Christo and full of talk of botanising, birds and holidays to come.

Back home, the euphoria of the Dixter visit disappeared quickly as Beth faced the work that needed to be done for Chelsea and around the garden. Andrew, she felt, was being unusually supportive of her concerns. 'We shall survive,' she wrote. 'I must make myself think positively as my American friends say, count my blessings and try to accept my failures!! (I'm not good at that!)' This was made much harder when Beth opened the post to find a letter to Andrew from Hans, putting into writing what she had already learned about his many infidelities. They had, it appeared, been going on for twenty years. She did not record whether she passed the letter on to Andrew. That night, she went to bed early but woke at 2 a.m. with her mind racing. For two hours, she read Jane Austen's *Sense and Sensibility* and eventually found it calmed her, especially the phrase, 'She had learned to govern her emotions.'

Beth also had to prepare for a second lecture tour to Canada and the United States in June and July 1984, immediately after Chelsea. However, this was something of a disappointment after the exhilaration of her east coast visit the previous year. It started badly. Landing in Vancouver on 29 June, there was the usual long walk through corridors to clearing and customs. On opening her suitcase, a female official took exception to Beth's large bag of seeds. In vain she hunted for her itinerary to prove the purpose of her visit. Finally Beth told her that her host, Ken Hadley, was out in the entrance to meet her. To Beth's amazement, the customs official knew him and agreed to call him in. However, no amount of persuasion would let them release the seeds without a certificate. Beth was furious, especially when the woman said she would let her take them back to England. 'They are my gift to Canada and the US,' ranted Beth. 'We left her and the seeds in disgust.'

BETH WAS A NATURAL WRITER and a ferocious editor and critic of her own work – not to mention that of others. She was looking forward to getting started on a proposal for a book to be called *Places for Plants*. This never materialised, but in the meantime, she would wake up in the night with an opening idea or good introductory sentence, though she annoyed herself by failing to sit up and write them down. With her conscience pricking her, she found that her mind was 'both a blank and a seething mass of subject matter that must eventually go into it'. On this rare occasion, Andrew had jotted down some notes for her, suggestions for several possible introductory ideas. She studied them but they were soon discarded. 'I liked a lot of them . . . but I could not accept them as they were disjointed and not in my language or style.' Faced with a sheet of blank paper, she eventually managed two sheets of foolscap in a morning. However, another book, *Plant Portraits*, published in 1985, developed from a series of articles she was commissioned to write for the the *Sunday Telegraph* magazine.

Beth works to prepare plants for display at Chelsea.

After a tough winter, Beth began sorting out plants for Chelsea in February 1985. She spent a day throwing out corpses – santolinas, teucriums – 'old friends now past . . . horrid smell from rotting plants'. A month later, she woke with nightmares about Chelsea. How was she going to fill 30 × 20 feet when she had lost all her big plants? She need not have worried. The stand that year was greeted with as much enthusiasm as ever. The large rheum once again made the journey to south-west London, along with the grey hostas and many old favourites which had survived the winter. Yet there was always a fresh feeling to her stands at a time when azaleas and rhododendrons still featured in so many show gardens. Garden designer John Brookes* wrote ecstatically: '[Her] exhibit is of course superb – I want it all, NOW.

She has a great eye for arranging and presenting her plants – some of the gardens outside could learn a thing or two from her, just on presentation.' The comments that she heard – 'so beautiful', 'such lovely groupings', 'best in show', 'my favourite', 'always come here first' – worked, she said, like a glass of champagne when one is feeling almost defeated by talking, taking orders, selling catalogues, giving advice and sharing enthusiasm.

Beth was understandably a little tense at the thought of entertaining legendary cook Elizabeth David, a friend from Benton End days, shortly after Chelsea. Beth remembered it being a very cold June day. She had arranged big jugs of Oriental poppies by the windows but had the wood burner going at the other end of the room. Sticking to her vegetarian principles, she prepared sorrel and spinach soup with cornmeal and cheese muffins. This was followed by sesame seed bread, green salad with rocket and a potato salad with green mayonnaise, slightly minted. The meal was rounded off by a cheese plate with oatmeal biscuits and strawberries and yoghurt. 'She really seemed to enjoy it,' Beth later noted with relief. She was also pleased that Andrew joined them for a glass of sherry and was 'his old charming, relaxed self – amazing! – And so good for me to see him like that – not aggressive or provocative.' David Wheeler,* who had brought Elizabeth David to see Beth that day, remembered things rather differently. Elizabeth had confided to him later that she would have preferred a couple of glasses of something French and white and perhaps something 'eggy' in line with her soon to be published collection of articles, *An Omelette and a Glass of Wine*. Of course, remembered Wheeler, she was far too polite to make such a comment and, after a tour of the gardens, the pair left with gifts of fresh produce from Beth's vegetable garden.

'Elizabeth was so modest,' Beth wrote after the visit, 'genuinely so it seemed. I hope we shall meet again.' They did, that November, on a return visit by Beth to Elizabeth's London flat. It did not disappoint Beth's image of what Elizabeth's home might be like, as they burrowed their way through dark book-stacked passages and stairs to the basement. It reminded her of Lett's kitchen, no space for anything, a table full of bottles, oil, wine and things such as a much-used wooden chopping board and a circular-handled chopping blade. Eventually, after tea and wine, they ate roasted goat's cheese, then oysters and Beth's chicory, mushrooms together with Italian cheese and a quince compote. Elizabeth was keen for Beth to stay –

Beth and Susanne Weber at Laufen for Countess von Stein-Zeppelin's eightieth birthday celebrations in 1985.

Beth felt she was rather lonely – but eventually she left, giving Elizabeth some pot pourri and a copy of *The Damp Garden*.

Earlier in 1985, she had been delighted to receive an invitation from her great friend Countess Helen von Stein-Zeppelin to the Countess's eightieth birthday celebrations at her home Meierhof, at Laufen near Basle. The family celebration for twenty guests was held the day after Beth arrived in the Countess's home, with the table laden with champagne and glittering silverware. Beth described the event in great detail in *Beth Chatto's Notebook*, published in 1988, drawing on her diary notes. Both in her diary and the book, the emphasis was on plants, with Beth closely inspecting the nursery famed for its perennial plants and the Countess's collection of irises. Other visits were organised for her while she was there. For instance she was taken by Susanne Weber, the Countess's companion, who had worked at the nursery since 1947, to visit the Botanic Gardens at Tübingen, south of Stuttgart.

Three days after the dinner, there was a large party for nursery staff and friends with over a hundred places laid on trestle tables under the vine-covered walls of the family home. Candles were lit along windowsills and lined the steps. Although language was a barrier, several guests managed to tell Beth – to her delight – that they had been to her garden. There was even a gift of plants for Beth during the evening, as a guest, Heinz Klose, gave her a pink double hellebore and a hosta. Later Beth retired to her room furnished with beautiful antique pictures and furniture. But it was the books by her bedside that impressed Beth the most: 'Lots of Gertrude J. and Mrs Earle!!'

Her final day at Laufen was also her own sixty-second birthday. Beth came down to breakfast to find the table had been garlanded with flowers just for her by Anka, a German student who had spent time with Beth at Elmstead, and that there were many small presents to take home, including a bottle of plum brandy made with Mirabelle plums. The main challenge was to pack all the plants she had been given by the Countess's nursery. Always a nervous moment, Beth strode passed the customs officials on her return to Britain with her bags stuffed full of plants. It was, she wrote later, the happiest of holidays.

On her return, Beth found that Hans was not the only neighbour that she had to deal with. Since the land on which White Barn Farm stood had been sold, it had been broken up between Hans and another farmer called Jennings, with pathways and rights of way becoming a legal minefield. Jennings now owned a strip of land bordering the garden and the nursery. In late January 1986, Beth received a letter saying that if she didn't pay him £4,000 for three-quarters of an acre, together with over £3,000 for his legal expenses, and also erect a strong boundary fence while still allowing him rights to dig drains across it, Beth would lose access to her wet nursery, which lay out of sight of the house. She was incandescent with rage at being, as she saw it, blackmailed into paying out far above the going rate for agricultural land, and poor land at that. 'Once again I resent the male sex,' she ranted in her diary that night. 'They are indolent and greedy. They don't rouse themselves to either think or work to improve the land and their position in it. It seems they envy me the cars of visitors they see coming to my business. Would they wish to change places? To work such long hours, seven days a week, to have the complicated responsibilities? They rush in with a tractor two or maybe three times a year and the Good Lord can do the rest. Urgh. I am feeling bitchy. But I must quieten down and think carefully how best to deal with the men.'

Two weeks later, it was Beth's husband who was on the receiving end of her wrath. 'Andrew makes me mad. He persistently washes up the few breakfast things which I have asked him repeatedly to leave for M – of course he always leaves his lunch tray for me to wash up . . . he will make coffee for the office all week but never on Saturday or Sunday when I am alone. Makes me miserable.' Something was festering inside Beth and three weeks later it erupted as Andrew set off to go for his regular trip to the pub. 'For forty years Andrew has only been interested in his ecological studies,' Beth exploded. 'Really what else has he done? Even gardening when we were young he did in fits and starts. Now he's too tired to be bothered even with his study. He has no little day-to-day activities of normal married life

Beth in the damp garden backed by the house and, high up, Andrew's 'eyrie' where he worked on his botanical studies.

Beth and an agile David Ward preparing the stand at Chelsea in 1985. 'My garden at Chelsea is the most exciting, frightening, absorbing thing I can do.'

to carry us both along! Why should he not have learnt to shop, cook simply, years ago! He has always expected to be waited on. The Pasha.'

Quite why Beth was so frustrated with Andrew's behaviour at this particular time is not clear. She was shouldering the responsibilities towards the nursery staff and the emotional pressure of her standing as a multi Gold Medal winning Chelsea exhibitor. Andrew, for his part, was doing what he had always done. Beth rarely mentioned the other side of his behaviour, as an adored father and grandfather. And while he may have frustrated her and sometimes also the nursery staff, they all had great respect and affection for him as a quiet, gently humorous man who was never openly disloyal to Beth. Only once did he admit to a long-serving staff member that Beth's long-standing relationship with Hans had felt like a divorce. But that was in the past. The challenge was now to establish a balance in their relationship – and that usually involved Andrew being the one to make the concessions.

As always, Dixter was a place of escape for Beth. Back there again in April 1986, she spent the first evening talking non-stop with Christo, later confessing to her diary that she loved the house, its decor without fad or fashion. On her return to Elmstead Market, she felt her home was small, with too much white paint. For a moment she missed the timeless atmosphere of Dixter, but she knew her mood would pass and that she would once again see her home in proper perspective. Her spirits revived and when Chelsea came around, she said that even the 'problem' of having too many plants had not bothered her.

During staging, the dry side of the stand – which, for some reason, she feared more than the damp and shady ones – came together easily and well. 'Somehow' ,she wrote later, 'just the right combinations came to hand, and it did make life easier to plunge in bark rather than stage in pots and boxes and conceal them with wraps.' For the first time, she had finished staging by 5 p.m., aided as always by her small group of helpers led by Madge Rowell and David Ward. 'He is gradually learning and gaining confidence,' wrote Beth that evening. 'He knows when it's wrong, not yet able to make major decisions on grouping, but is getting the idea round the edges, of contrast, and background plants.' She was delighted when Graham Stuart Thomas arrived to walk around the stand, helping Beth to check names. She was even more delighted with his comments: 'full marks, fuller than full'.

The next day saw the usual parade of royal guests. Prince Michael of Kent was the first visitor (no Princess Michael this year), then Beth met Roy Strong, then director of the Victoria & Albert Museum. Next, Princess Margaret, looking, Beth thought, much better, slimmer and happier than she expected, and animated. They walked around the stand with Roddy Llewellyn, the landscape gardener who was also the Princess's lover. They all chatted informally, with Princess Margaret commenting, 'Isn't she helpful?'

Next came Sarah Ferguson, who had become engaged to the Queen's second son, Prince Andrew, in March that year. Later to become renowned for her exuberant behaviour, she had, she said, just been dazzled by a hundred vases of roses, so Beth was not surprised that she hardly registered her 'little subtleties'. Finally came Princess Anne, looking, Beth thought, beautiful, with lovely skin and eyes and perfect teeth. They chatted for several minutes and Beth found her very warm and human. She admitted to Beth that her own garden was really wild, as she had little time for it. Beth told her she was too busy doing things for others but that, as one gets older, gardening is a way to go on being creative as one cannot go on having babies. Princess Anne laughed and said she would remember. It was an aphorism that Beth was to repeat many times over.

Later that year, Beth had a similar experience with the Queen Mother at the Sandringham Flower Show. Although the Queen Mother had a reputation for being interested in gardening, Beth felt she showed she was out of her depth when she called a *Phygelius* 'Yellow Trumpet', a lovely 'primula'. The Duchess of Kent revealed similar ignorance, Beth thought, by regretting the plainness of hostas but loving the 'ferns' which were, in fact, cherry-red astilbes.

A while back, Lambeth Palace, London residence of the Archbishop of Canterbury, had approached Beth to ask her to design a border for the gardens there, the biggest central London gardens after Buckingham Palace. Beth took the opportunity of Chelsea week to visit to see what she thought. She was greeted and shown round by the then Archbishop, Robert Runcie, whom she mysteriously described as 'delicately perfumed', and decided that she could supply plants for a border 90 yards long by 4 feet wide. She did later supply a list of suitable plants, which remains in the Palace's archives, although the border has since been replanted.

Beth and the Queen Mother at the Royal Sandringham Flower Show in 1986.

With the stress of Chelsea and all these demands on her time behind her, Beth again concentrated on her own garden. There was planting to be done by the reservoir between the taxodiums and weeping willow. Trolleys were loaded with dark purple-leaved cannas and variegated *Arundo donax*. Beth felt this bed was her one piece of extravagance with exotic plants that would have to be brought in in winter. She added border phlox, zantedeschia, some filipendula to go with two tall miscanthus at the back. There were more purple leaves with *Ligularia* 'Desdemona', and the red-leaved grass, *Imperata cylindrica* 'Blood Grass', she had been given by Marshall Olbrich.* Several hundred cuttings of *Euphorbia palustris* and *E. myrsinites* were taken since Beth knew they were both difficult to strike.

BETH'S SUMMER HOLIDAYS with Hans had now been replaced by visits to stay with Helen von Stein-Zeppelin, and once again, in July, she flew to Basle and on to Laufen. This visit, now on her own, was, as usual, the ultimate busman's holiday. Beth spent hours in the nursery helping with potting up and studying those plants that were unknown to her. She was attracted by practical matters as well, doing a

Beth was never sentimental about animals and Emma, the nursery cat, had to earn her keep catching mice.

sketch of a little set of shelves to have made for the work bench in Elmstead Market, to hold labels, blades, knives, secateurs and the like, with below them a place for work sheets and record cards, and at the top a row of nails to hold the roses from cans or boxes.

In the evening, she retired to her bedroom with its collection of books, continuing her reading of Mrs Earle's works that had belonged to the Countess's mother. She enjoyed them enormously, particularly when she came to the section on health and diet. Beth was delighted to read that Mrs Earle had become an enthusiastic vegetarian, believing that uric acid forming foods were the prime cause of ill health (apart from germ infection). Mrs Earle said that it would take generations for this to become accepted and Beth agreed, pointing out in her diary that vegetarianism had become much more widespread in the 1980s but, she wrote somewhat disapprovingly, 'often for sentimental reasons rather than any consideration of the effect on health'.

Now sixty-three, Beth was conscious of all her regular aches and pains, whether in her neck, back or hands, and regularly sought help from an alternative therapist in Ipswich for massage and manipulation. His insistence that Beth try and relax more was hard for her to accept. 'I know I should try to relax more,' she admitted to her diary after one of her treatment visits. 'I don't find it easy. Partly because there is so much I want to do. My work is my pleasure, not a duty. But sometimes I know I just cannot relax even if there was nothing to do. I need to do something – to be happy with myself. To stop feeling lonely – But it is not just that. I am driven to do things, not just to tidy, collect, record answers – but to create something – a meal, a bowl of flowers – set the atmosphere in a room with pretty living things around me – to create the whole broad canvas of the garden and nursery.'

BUT THERE WAS NO LET-UP in the relentless round of visits, talks and lectures, culminating that autumn in another trip to the United States, where her international reputation was now assured. The trigger was a plant symposium held over ten days at four locations starting at the New York Botanical Garden, followed by Chicago, Washington, DC and Boston. Although it involved an exhausting schedule, being asked to speak at the symposium was both an honour and irresistible to Beth since it gave her the chance to meet up with many in her new circle of American horticulturist friends such as Larry Pardue and James van Sweden.*

Her lecture was on 'Design Throughout the Year'. She was well equipped, with eighty slides, divided into sections, starting with 'First Signs of Life', with early spring delights such as galanthus, hellebores and, of course, her beloved bergenias. She then moved on to views of the 'Mediterranean Garden', followed by 'Paved Areas, Steps and Courtyards', 'Cooler Areas with Retentive Soil'; a long section on the 'Water Gardens' followed by the 'Reservoir Garden on Clay'; and lastly, 'The Year's End in Sight', with final shots of a hoar frost to emphasise the form and texture of shrubs and trees. Although the standard of all the speakers was high, it was Beth who was given a standing ovation. With an audience of over five hundred, she was later besieged with people asking her to sign books.

However, she learned from the other delegates, noting on her symposium programme during a talk that she must remember to pace her speech since a hammering voice, relentlessly pouring out facts, is very tiring. 'Say less,' she scribbled, 'it will mean more, and give the audience some respite.' She was particularly impressed by J. C. Raulson from the North Carolina State University, making a note to herself to point out that her garden was a living catalogue showing what plants would be like with six or seven years of growth. But it was James van Sweden's talk on 'The New American Garden Style' that prompted Beth to write the most notes, from 'tulips in spring used as annuals, coming through grasses and herbs' to 'must put more tall grasses with achilleas – *Stipa arundinacea*, *Stipa gigantea*.'

In between the four lectures Beth was there to give, a whirlwind itinerary of guest and garden visits had again been arranged, with enthusiastic hosts always anxious to show Beth their gardens and organise entertainments for her. She was grateful, therefore, to see that a brief break had been built in for her to spend a short time at the home of Paul Aden,* whom she knew briefly from her previous US trip and a visit he had made to the gardens at Elmstead Market.

Aden picked Beth up from the office of Rosemary Kern, the symposium organiser, to drive her to his home on Long Island. He was, Beth described in her travel notebook, 'a grizzled bear-like figure, wearing a blue cotton fishing type hat'. The journey was relaxed and they reminisced about the great hosta grower Eric Smith, and chatted about whether they should together write an appreciation of Eric for the RHS's journal, *The Garden*. Half an hour later, they arrived at Aden's home, a typical white-painted boarded house, and after dumping Beth's bags, went straight into the garden full of hostas, grown for evaluation to see how stable the variegation will be or how large the plants will grow. Beth knew that young or newly planted hostas looked quite different from mature plants. There were also many shade-loving plants tucked away among them, mostly variegated but not all. He had unusual epimediums, different grasses, unusual forms of the Japanese painted fern, and various larger forms of a green and white fern. Beth thought that many of the hostas were very fine, but some too similar to make her want them. But a few outstanding ones she thought were possibly useful for edging or a raised cool bed, while one extremely large, particularly firm, as yet unnamed plant with shining plain green leaves, oval with long pointed tips, struck her as very beautiful.

Eventually they stopped and sat on a seat under an open-ended structure and talked at length about all kinds of things, but particularly of writing. Paul was writing, with others, a book on perennials but finding it hard to source pictures of

Beth examining hostas during her trip to the east coast of the US in 1986.

good combinations. He was full of compliments about Beth's garden, and said it had been almost a religious experience to see it on his last visit the England. '"There now!", as Cedric would have said,' Beth wrote in her notebook that evening. She was touched by all the nice things he said and told him she was still influenced by Andrew's ideas although he did not always approve of her interpretation of them. But, as she explained to Paul, she would not have known how to design a garden in this way if it had not been for Andrew. Paul said it was original and rare to manage not only to make the garden but also develop the nursery, collect and expand the collection of rare plants – and make it pay. He talked about the big nurseries and some of the problems that go with selling in large quantities – not always profitable

– with buyers always wanting the lowest price and the name of the producer lost before it reaches the customer. Beth enjoyed getting to know Paul without what he described as 'the crutch of plants', much as they both loved them.

They then spent a while looking at Paul's slides of his hostas. Beth thought he had made some lovely ones but was relieved that he seemed loath to name and introduce any that were not exceptional. They then descended to his cellar where he kept all sorts of intriguing equipment including propagating shelves. But Beth was particularly fascinated by his computer – a cutting edge and still exotic Apple Macintosh. Paul let Beth have a go and although at first she was bewildered by it, slowly she began to follow what it could do, quickly realising that the more effort one put into it, the more it would do for one.

After lunch, Beth went back to the computer to read chapters of Paul's new book on hostas and perennials. She was delighted to have progressed sufficiently to sit at the keyboard and turn the pages, even learning to correct spelling mistakes. Paul's writing, she found, was very good, condensed but well thought out. She was very impressed.

Two days later, after more garden visiting, Beth and Paul had a Chinese meal on the way back to his home. Once again, they spent the evening talking about plants and the possibility of Beth propagating some of Paul's rare hostas. Then Paul started to talk about his marriage. It was clearly unhappy, and he told Beth he and his wife were leading separate lives. She recalled trying to help by being sympathetic, since she liked them both. But the next morning, after a walk around the garden, Paul embraced Beth, saying they must be lovers and that he was planning to buy a house in England. She genuinely does not seem to have seen this coming. 'God spare me!' she wrote in her travel notebook that evening. While she felt sad for him, 'it would solve nothing and I have enough problems!'

In contrast to the excitement of New York, Chicago was calmer. The speakers stayed in a hotel and Beth was able to have time getting to know them better and meet the new speakers who joined the group, including garden designer and historian Penelope Hobhouse* and Allen Paterson, who was at that time Director of the Royal Botanical Gardens, Hamilton, Ontario. Then it was on to Washington, DC, where Beth, Penelope Hobhouse and her husband, John Malins, and Allen Paterson, went off to visit the National Arboretum. Disappointingly, it poured with rain but the group still managed to see the Herbarium and the bonsai collection. They also came across a collection of old roses including a 'Champneys Pink Cluster' in bloom. Paterson, an expert on old roses, pointed out 'fastidiously', as Beth later noted, that it smelled of Pond's Cold Cream. He would doubtless also have told them that it was one of the first transatlantic roses, having been bred from seed sent from France in the late eighteenth century to Charleston, South Carolina, and later returned to Paris to become the parent of French Noisette roses, named for the two Noisette brothers, each of whom had a nursery, one on each side of the Atlantic.

By this stage of the trip nothing would have surprised Beth, having been entertained in Washington by a hostess – 'not short,' Beth noted, 'it would appear, of a dollar' – whose late dachshund used to have his own seat when she flew on Concorde. Beth was able to visit Dumbarton Oaks near Washington. She had been told there would be nothing of interest for her there. However, she loved the atmosphere – especially precious since she was almost alone there, an unheard-of luxury on these trips.

The final stop of the trip was Boston, which immediately felt the most European to Beth with its covered walkways and market stalls. But there was little time for shopping, as the group was whisked off to the Arnold Arboretum for the final set of talks. At the farewell dinner later that night, the director of the arboretum came up to Beth and told her that 'they' said she was the best speaker. 'Well,' Beth replied, 'perhaps different.'

The next day, the group said their goodbyes and Beth travelled by Greyhound bus north to White River Junction in Vermont to stay for a few days with Lucie S, a gardening friend, before flying back to England on 23 October. She was met by her estate manager, Keith Page, at the airport and was soon back home at White Barn House, unpacking her bags, washing clothes and sorting papers. But first she dashed to her vegetable garden, picked the remains of the climbing French beans, some calabrese, and a Chinese cabbage . . . 'So very good to be home!'

9
Annus mirabilis

WITH A TOUCH OF NONCHALANCE, Nigel Colborn reported in May 1987 that 'Beth Chatto earned her usual Gold with her unusual plants skilfully juxtaposed for harmony and contrast . . . After the razzle dazzle of the geraniums and busy lizzies, it was a relief to rest the eyes on some of the hardy perennial displays.'

But the year had started with disappointments, though these were felt more by other people than by Beth herself. In January, Valerie Finnis had called to say how angry and upset she was that Beth had not been awarded the Victoria Medal of Honour by the RHS. This was their highest honour, named for Queen Victoria and only ever held by sixty-three leading horticulturists to celebrate the number of years she reigned. Beth was more stoical. The awards that year had not been to plantsmen and, anyway, Beth had always been clear that she was not a servant of the RHS. She had refused twice to go on their committees. It was the friends who shared and loved the garden, she insisted, who really mattered to her.

Christopher Lloyd was also cross about her omission from the VMH honour list. When Beth protested that it didn't really matter, he insisted that it did to him – he had been awarded it in 1979 – and to many of their friends who would have been thrilled if she had got it. 'Well,' said Beth, 'that is enough for me. The friends who share and love the garden give me so much more.'

By the end of March, she finally finished writing the *Notebook*. It had taken her much longer than her previous two books and it was a relief to send it off, releasing her to start serious work on that year's Chelsea stand. On the first day of setting up in mid-May, Beth started work early, at 5.30 a.m., as she always did at this time of year. Going out into the wet garden to cut stems for the stand, she was concerned that she had no large *Euphorbia wulfenii* plants and made a note to grow some for the following year. This year she would have to make do with some tall flowering stems cut fresh that morning. She also cut some flowers of *Asphodelus albus* to arrange with pots of leaves without flowers, and a handful of *Convallaria majalis*, lily of the valley, to add to the little pot plants already packed into the van for the journey to London.

At the Chelsea site, David Ward was now the main stalwart at Beth's side as they began the staging, 'such a good companion and assistant on these occasions', she wrote in her diary. 'We shared our lunch baskets, his home-made bread, cheese and mini sandwiches, my radishes and avocado.'

While the Chelsea marquee has always had a well-deserved reputation for a sense of camaraderie among exhibitors, it was only human nature for stand-holders to eye up their neighbours' exhibits. 'I have brought more grasses this year, I am determined to use them to effect,' Beth wrote in her diary. With her old friends

Beth at Chelsea with her team (left to right) Harold and Madge Rowell, Janet Crosby, David Ward and Lesley Hills.

Blooms of Bressingham right beside her stand, Beth was well aware that they were exhibiting many of the same plants. She therefore felt she must do something different. 'I haven't seen a single grass yet among their collection,' she noted with satisfaction, 'but their staging of perennials is much improved (*à la* Chatto).'

Ever the perfectionist, Beth would often change her mind about arrangements. A bulky mass of hostas staged the day before was pulled apart and lightened with verticals such foxgloves and aquilegias. By the evening, they had made a group on both ends, staging the tall and bulky things, concealing the boxes with the tall grass *Arundo donax* var. *versicolor* and several tall miscanthus.

On the final Sunday, as always, there was curiosity from other plantsmen to see what Beth was exhibiting. She was visited by Joe Elliott, who ran an alpine nursery in the Cotswolds. He told Beth she had done more than anyone in the world to influence the showing of plants at Chelsea. Beth replied that she could not be compared to the big commercial nurseries such as Hillier's, Notcutts and others who had been doing it for fifty years or more. Elliott replied that it was not the same at all, as Beth had done it as an individual.

As regular visitors to Chelsea will know, the weather in the third week of May is notoriously unreliable. In 1987, it was particularly cold and Beth had to adapt her outfit for the traditional royal visit on Monday afternoon. Rushing back to the Terstan, she got out her new lightweight summer suit, blue and covered with tiny flowers, and a Nile green jacket that matched the green in the floral pattern. But first she put on two Damart thermal vests and a long petticoat in a bid to keep warm during the inevitable standing about that afternoon and in what turned out to be a 'perishing cold evening'. This year, Beth saw the Queen and the Duchess of York only passing by but was presented once again to Princess Anne. 'She was charming – so natural and easy,' wrote Beth that evening, 'but knows little about gardening.'

The next day Beth arrived at the marquee to find yet another Gold Medal; she had no idea at this point that it was to be her last at Chelsea. She was delighted and amused by some of the comments she received. Lord Aberconway, past President of the RHS, who oversaw the famous flower show, as usual had told Valerie Finnis and Rosalind Runcie, the Archbishop of Canterbury's wife, that he had watched her place every plant – not true. Mark Rumary of Notcutts told her she had created 'a landscape with herbaceous plants'– that it was magic.

In summing up her Chelsea exhibit for 1987, she was happy that so many people had made similar comments. 'It is thrilling, heart-warming to hear, after all the weeks of preparation and anxiety – not knowing at all how it would turn out until I actually started to build my Chelsea garden. I think this has been the best design I have made yet. It had gentle flowing lines, spaces beneath and around plants making shadows, or showing the shape of stems as well as leaves or flowers. It has been the happiest Chelsea.'

At the end of the flower show week, Beth and her team returned to Elmstead Market to try and catch up on the garden maintenance and nursery orders. She was understandably exhausted, and it took her a few days to recover before she could

Beth and helper Caroline Griffiths working at the propagating table.

such a style, but I admired very much many of the mature combinations especially of clematis. Wonderful to have so many tall walls to clothe but clever to know how to clothe them. I enjoyed the morning much more than I expected.'

In contrast, Beth and Christo's trip to the RHS gardens at Wisley in Surrey was less of a success. 'I find Wisley a strange place,' she wrote in her diary that evening. 'There is nothing there now to thrill or excite me. I don't think it's because I've become blasé – there were plenty of lovely ideas at Sissinghurst and all beautifully, lovingly planted and cared for. Wisley does not have that look. It feels like a public park. On the whole the plants are chosen for public consumption. I thought the main herbaceous borders were crude and clumsy – not any group that could be called attractive – just blobs and blocks of plants with little care, it seemed, of setting each other off. It is easy to be critical. I did say [to Christo] we should try to imagine what we would do to improve it. Add more shrubs. The backing of a clipped hedge with no more trees or shrubs is very boring. What must it look like in winter?'

SEPTEMBER brought another Gold Medal at the RHS's Great Autumn Show. When, on 17 September, an envelope arrived with the familiar RHS stamp upon it, that was no great surprise. However, this one was different. It was from the President of the RHS, Robin Herbert, inviting Beth – finally, as some of her friends would say – to accept the Society's highest award, the Victoria Medal of Honour. It was, Herbert wrote to Beth, 'fitting recognition for all that you have done to inform and inspire gardeners through your writings and by the example of your nursery and especially through your splendid exhibits'. Beth did not reply immediately. She was overjoyed by the news but also, as always, concerned not only for her own reputation but also for that of the gardens and nursery. Six days later, she wrote her acceptance letter, with an additional request: 'I am happy and proud, that you and members of the Council feel that I am worthy of this coveted award. But, as every season makes me aware of the ephemeral nature of gardens, and gardening, I hope I may continue to be worthy of it. If your letter were to be quoted . . . would you kindly consider adding "by the example of your garden and nursery". Without the garden there would have been no nursery and now the two are essential to each other.'

Beth in one of the tunnels with helpers Jo Enever and Sue Taylor.

Beth was sworn to secrecy about the award until the official announcement in early December 1987, so it was with relief that in the middle of October she was able to get away for a short break to Germany to stay again with Countess Helen von Stein-Zeppelin. The visit to Laufen was a last-minute addition to a planned trip to Holland for a speaking engagement organised by Romke van de Kaa, Christo's former head gardener and now a leading nurseryman and horticultural journalist. It was to be Beth's first talk in continental Europe. The Countess had called her ten days before, saying she had seen an article about Beth in *The Garden* and asked when she was coming over. Although Laufen was in the south of Germany, and so out of her way to Holland, she was always anxious to see her friend, who had become something of a mother figure to Beth since her own mother's death.

The trip started well, with Beth spending hours in the nursery taking notes on their packing methods and recipes for compost. However, on the night of 15–16 October, the East Anglian coast was hit by the worst storm in two hundred years. Having flattened huge swathes of the Home Counties, the 'Great Storm of 1987' moved across Essex, Suffolk and Norfolk, where gusts of over 120 miles per hour were recorded. The next day, once it was safe to go outside, with Beth in Germany, it was up to Andrew and David Ward to assess the damage. While the destruction was not as bad as in some parts of the country, the woodland area had suffered badly. This was especially painful for Andrew, who had deliberately kept this area of land as an ecological haven for wildlife. Now many of the tall trees lay horizontal in mangled heaps of snapped branches and exposed roots. Among many losses elsewhere in the garden, the crack willow had been ripped out of the soil. Helpless in Germany at Laufen, Beth could only listen to radio reports of the extent of the damage and see on television the devastation wrought across the south-east of England. With the phone lines down, there was no way to speak to anyone in Elmstead Market. She was left with no alternative but to make the 500-mile journey to Holland as planned the next day.

ON HER RETURN TO ENGLAND after her speaking commitments in Holland, she reflected on the devastation in her diary. 'It took me most of the day to . . . take in the storm damage. Every time I go round I see something I hadn't seen before. It could have been worse I know. The framework of the garden still stands – the Big Oak lost a few branches, but stands, defiant, on the headland. I can't imagine the garden without it. It is still the finest thing here. Stretching so far back into the past, having weathered so many storms – wars – and all kinds of perils – it gives me strength somehow – I can't bear to think of the garden without it.'

While the garden had not suffered as badly as some areas of Suffolk just to the north, where vast swathes of the Rendlesham Forest had snapped like matchsticks, the damage at White Barn House was enough to force Beth to reconsider her plans. At the beginning of November 1987, with mixed emotions, she wrote to John Cowell, Secretary to the RHS, informing him that 'with regret' she would not be

able to show at Chelsea the following year. She gave her reasons as both the long spell of eleven years' showing, taking more than four months of her time every spring, and the effects of the hurricane, which had left the garden well below the standard she hoped to achieve. He wrote back immediately commiserating over the devastation with the suggestion of an opportunity to relandscape for the future and reassuring that space would be found for her in 1989.

It was not to be. Perhaps with a sense of prescience, John Cowell wrote to Beth again at end of November to inform her that the RHS Council had awarded her stand at that year's Chelsea the Lawrence Medal for best exhibit, adding 'how we will miss a 1988 exhibit!' A delighted Beth replied saying that she was astonished, thinking it courageous of the Council when not so many years ago, 'our plants (species) were considered weeds.' She wrote in her diary that night that she was quite shaken, totally surprised and elated. 'When I come to think about all the other great efforts that are made by prestigious firms of long standing . . . well, as Cedric would have said, "there now!"' After lunch, she went out and potted until she ran out of space. 'It felt great to be planning and preparing for next season's pot garden. I still have the geranium Cedric brought home from the Canary Islands, and his special large grey succulent . . . I told Andrew and Seley about the letter but must wait till the announcement is made in the press.'

There were distractions in the form of a dispute with Dent, the publisher of her first two books, who claimed that an exclusivity clause should stop the publication of *The Green Tapestry* by Collins, already at mock-up stage. Beth was horrified that Dent, worried that she was becoming famous through another publisher, were insisting that all her future books should be offered to them first. When she had signed her contracts with them, she did not have an agent who would have picked up on such restrictive clauses. Now, on Christo's recommendation, she signed with his agent, Giles Gordon, who was doing his best to defuse the situation. Nevertheless, it infuriated Beth and she fought hard to have the clause removed, particularly since she claimed that at least ten major publishing houses – including, she noted indignantly, Chatto & Windus, Andrew's old family firm – had asked her to write for them. 'I do not like the idea of being "hot property", feeling I am being fought over like dogs with a bone! It puts too much onus on me to go on producing something special. Right now I don't feel capable or enthusiastic enough to write anything.'

She also had to visit her accountant, who wanted to discuss the future of the business and, with the tact required of good accountants and lawyers, asked her about the possibility of retiring. She had still not calmed down by the time she wrote in her diary that evening, 'If I am useless and senile then I would be best out of it. But for the present, and I hope several more years yet, this is my life. What I would do without the garden and nursery I cannot bear to contemplate. Every day I am glad to wake up knowing there is more to be done that I love to do – than I can possibly do. As for a manager – the perfect person might step in one day – but hasn't yet.' The truth was that Beth was not looking for anyone to be her successor and her choice of words is revealing. David Ward was successfully overseeing the propagation and nursery while Beth was always 'head gardener'.

But the meeting with the accountant did give Beth pause for thought. 'After Christmas, Andrew and I must arrange to meet our accountant and solicitor to remake our wills. There are so many things to consider. I want to look after my daughters and their families, but I also feel responsible for my staff, especially the younger ones who will have a future after me. I also would like to make provision for some really worthwhile cause . . . I do not feel it would be wise or good for my grandchildren to inherit too much. They must be encouraged – but also have the incentives to struggle for themselves.'

December brought a letter from Isbert Preussler, Countess Zeppelin's gardener, along with an English translation. 'I found it very touching, he feels much the same about me as I do about him,' she wrote in her diary on the last day of this personally momentous year for Beth. 'We would both have so much to say to each other if only we had a common language. In a sense we have the love of plants and people too. He compares me with Karl Foerster. That is, I know, a big compliment. I feel so lucky to have these warm friends.'

The end of a year always made Beth reflect on its highs and lows. Once again, the darkness of winter brought on feelings of depression. Her outstanding achievements do not seem to have brought her great comfort. 'Sometimes I long for a kindred spirit to be near to me when I am alone – usually the evenings – or during the day when the nursery is closed and the staff are not here. When they are around I am never alone, they are kindred spirits about the place. Sometimes I feel it is bad, a weakness, to have such a strong need for close companionship. I do have many good and sincere friends and am elated by their company, sharing pleasure (or worries) with them. I can be content alone. So long as I am busy, either physically or mentally – but I tend to become depressed if I am alone without the stimulus of some interest.'

When the news of both Beth's awards from the RHS was finally made public, she was inundated with congratulations. With her usual desire for order, she kept a list of names of those who wrote or telephoned with brief biographical notes. They ranged from horticultural luminaries and old friends such as Graham Stuart Thomas and Peter Seabrook (who arrived in person with a bottle of champagne) to the more prosaic letter of congratulations from her bank manager. The first card to arrive came from Valerie Finnis, who had been awarded the same medal in 1976, summing up the feelings of many in the horticultural world: 'Beth Chatto VMH at last! . . . Heartfelt congratulations, a long, long overdue recognition of your enormous contribution to gardens and gardening the world over.' Another came from Kath Dryden, President of the Alpine Garden Society, also a VMH holder, 'Well done, well deserved – Up the girls is WOT I say! Live long and enjoy your VMH.' Philosophically Beth noted, 'It is the warmth and characters of all my dear friends sharing the pleasure with me that makes it happy. Would be deadly dull and lonely by oneself.'

The medals were to be awarded at the RHS's AGM, with a celebratory dinner for holders of the VMH the evening before. Beth had hoped to stay with Valerie Finnis that night but her London flat was too small, so Rosie, Beth's secretary, booked her into the old Chelsea favourite, the Terstan in Earls Court. The next challenge was to decide what to wear.

Although Beth feigned a dislike of shopping for clothes or spending money on them, she was always very conscious of what she wore and wrote about it at length in her diaries, which are littered with descriptions of frustrating shopping trips to Colchester, or just occasionally to London, while she searched for just the right item to go with something in her existing wardrobe. Around this time, one such trip had taken her into the lingerie section of a local department store. 'I saw delicious undies with really good lace were £35 for the camisole top, and

£25 for the frilly pants. Decided it was absurd to want such things at my age . . . but, of course, I would have loved to buy and wear them.'

However, there was to be no frivolity for these two events. Although Rosie tried to persuade her to go out and buy something new, Beth instead turned to her existing wardrobe for inspiration. Like so many women, she had her favourite outfits that came out regularly and helped her feel relaxed. For the awards ceremony, she decided to wear a woollen Chelsea suit she had had for several years. While she no longer did any dressmaking herself, she often had pieces altered by a local seamstress, always aiming for a perfect fit. The suit's skirt had been shortened since she had bought it but although she was wearing longer skirts now, she decided the balance was right with its short boxy jacket. However, she rejected the pale fawn chequered blouse that she'd previously worn with it as too safe and insipid and decided to go for a new blouse bought recently in a sale which was of a rich arabesque pattern in Indian reds and peacock blue. Gloves, stockings and bag were all chosen to complete the outfit. She showed Andrew, who approved. For the dinner, she chose another old favourite, what she called her 'Quaker's dress' from Laura Ashley, long with a large white collar, together with a red shawl and a pair of new earrings made in Thailand.

While Beth never enjoyed being in front of the camera – an enigmatic smile became her trademark look – she was the consummate professional, always remembering what she had been worn before for a repeat shoot, with the same scarf, suit, brooch or blouse reappearing. She was adamant that while she enjoyed people listening to her 'prattling', she hated being looked at intently, being the focus of attention whether in front of the camera with a photographer or, on one occasion, an artist, Richard Stone, who painted her portrait. She prepared for this with unexpected zeal. 'I put slices of raw potato under my eyes, wistfully hoping they might reduce a little of the puffiness. I hate the bags under my eyes. I wouldn't mind so much if they were lines (I wonder if I will ever be bothered enough to have them removed – it must take courage to go to a doctor to ask). But more difficult to "remove" is my nervousness, self-consciousness, all that nonsense that makes me look unlike my true self – that put on for the photographer look!' She later did have her eyes 'done' and went to stay with her daughter Mary until the bruising had subsided.

Beth was the only recipient of the VMH in 1987, and the only person to receive both the VMH and the Lawrence Medal in the same year. She sat in the place of

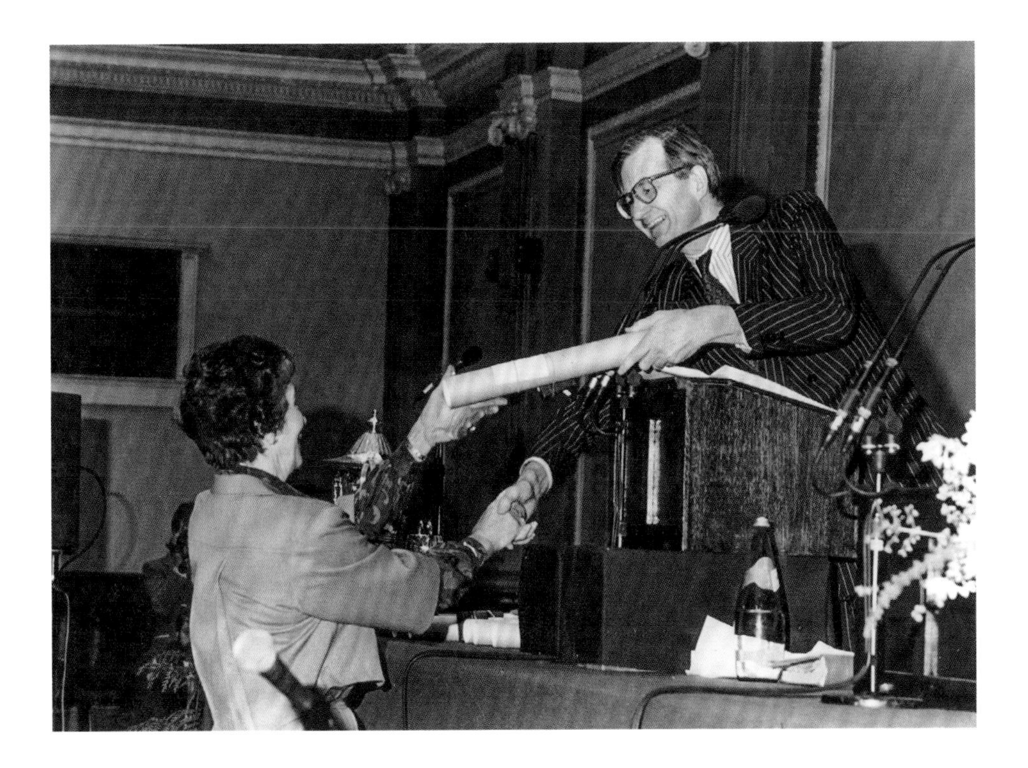

Beth receiving the RHS Victoria Medal of Honour from Robin Herbert in February 1988. She was also awarded the Lawrence Medal for the best exhibit throughout the current year's shows.

honour between the RHS's President, Robin Herbert, and John Bond, Keeper of Windsor Great Park and the Savill Gardens. Sitting on Robin Herbert's other side was the famously terrifying Princess Sturdza, whose garden, Le Vasterival, near Dieppe, Beth had visited on a trip to France with Hans in 1980. Despite a delicious menu of spinach roulade, escalope of veal and cinnamon biscuit cake accompanied by Mâcon-Vinzelles wine, Beth had no appetite. Such was her excitement, she reported, that 'I did not really give the menu my proper attention.' The menu card itself however did remain a treasured memento of her evening of glory.

10
After the storm

WHILE MOST of Beth's friends and colleagues in the horticultural world were disappointed by her decision to stop doing Chelsea, some understood completely. The author Penelope Mortimer, a keen amateur gardener, had become a friend and she wrote to Beth about her decision with glee. 'How wise, how very wise to abandon Chelsea. I'm absurdly pleased, a) that someone as distinguished and as much in demand as yourself has the courage to cock a snook at the Establishment, and b) because I now needn't go there again. I've always hated it – and the work and hassle for you must have been unbearable.' Although Beth herself did not see it as 'cocking a snook' at the RHS, being no longer chained to the Chelsea Flower Show routine had its rewards.

Nevertheless life showed no signs of slowing up in 1988. It was to be a year of more honours, holidays, writing, radio appearances and, of course, gardening. Beth was also finally able once again to travel for pleasure and she wanted to reassess the gardens. After the devastating hurricane of October 1987, she had plans for the woodland area. There were constant requests for talks and appearances, books to work on and articles to write. She was delighted that Andrew was finally willing to give her some ideas for the *Places for Plants* book project. She heard him 'banging away in his room on his old typewriter'. 'It will be exciting if we can do something together,' she wrote that evening. 'And good for him to be working, putting into print some of the principles and ideas that I have based my planting on over 40 years.'

Beth always hated winter and reacted badly to the cold and the low light levels. On a mid-January Saturday, she switched on the electric blanket in the spare room and huddled under it 'with pleasure – didn't sleep but tried to relax – deep breathing and all that'. It had taken her all morning to write six hundred words on

'Gardening in Shade' for the American journal *Horticulture*, finding it difficult to pitch herself using so few words. Restored, she went out into the garden, where it was drizzling lightly, for some fresh air, stooping to pick a handful of the small, delicate *Narcissus* 'Cedric Morris' to revive the displays inside the house. She told the story ten years later in *Dear Friend and Gardener* of how a single bulb had been found in Spain in the late 1950s by Cedric, who later gave some to Beth and which she had spent years bulking up.

But this was only a short distraction from the seemingly endless writing projects she was involved with, including *The Green Tapestry*. The plan was that the text for *The Green Tapestry* would be edited from recordings made with its editor, Susan Berry.

With no Chelsea to plan for, Beth had more time to devote to the gardens.

Over several weeks, Berry arrived at the garden armed with a tape recorder and walked through the various sections. It was a fraught relationship as Beth did not like this way of working and never felt that the book spoke for her as her earlier ones did. 'I always feel apprehensive before she comes, wondering how on earth I can turn on enthusiasm and inspired ideas or even practicalities about the garden from cold,' she wrote that January. However, by this stage she liked Berry, and she was impressed with the layout and photographs by Steven Wooster, one of her favourite photographers.

At the same time, she was being visited by Clare Roberts, who was working on the line drawings for what was to become *Beth Chatto's Garden Notebook*.

After years of handling plants, Beth was increasingly bothered by skin allergies and wore gloves when gardening.

Beth loved Roberts's work: 'It is quite breathtakingly beautiful, such sensitive, exquisite drawing. So often her designs exactly capturing my principles, the feeling for form, texture and the "golden triangles". I hugged her with gratitude and admiration.'

Shortly after, she had the chance to celebrate and remember her magical trip to Château Mouton Rothschild. One evening, Hans collected her to have supper with him. He had a surprise for her. Having noticed that Baron Philippe had died earlier in the week, he had bought a bottle of Mouton Cadet in the local supermarket and they opened it that evening in memory of the Baron, washing down a special cheese from the Auvergne. Beth was touched and the evening, she wrote, a happy one, remembering the Baron, a man who, she said, was 'naughty [but with] great charm'.

Later that January, her post contained the usual correspondences from gardening friends. John Dyke wrote to her on behalf of all Beth's friends at Notcutts Nursery, complimenting her on being the first person to have been awarded the VMH and the Lawrence Medal in the same year. Graham Stuart Thomas wrote to suggest a name for a variegated *Pulmonaria rubra* that had been discovered at the nursery. While Beth put forward 'Chatto's Cream Edge', it was eventually to be named 'David Ward' for her nursery manager, who had found it. Another letter came from Essex University saying they wished to confer an Honorary Doctorate on Beth. She was, she wrote that night, 'struck dumb' that they should want to give her such an honour.

It was not all good news however. In March, Beth went into the packing house and commented on a lovely box of plants that they were packing. She was delighted to see they were for the RHS's gardens at Wisley. However, a few weeks later, she received a letter from the Rock Garden Superintendent complaining of the poor quality of the plants, that several were dead and the rest needing special care if they would survive at all. It transpired that they had been stuck in the post for a week over Easter, with dire results. Mortified, Beth wrote back apologising and saying they would replace anything if the stock was available. Stock control was always a problem but dealing with the post office for deliveries even more so.

Beth was going through one of her troughs of depression. In her mind, the garden was being too busy. 'I should be pleased, but seeing children dashing over the poor grass, women pushing prams and pushchairs – ugh – it has become a pleasure park,' she wrote despairingly in her diary. 'Have the best years gone? Have I lost my garden for money at the gate – I feel very depressed. Yet I could not have had such a garden without the nursery and the staff and where is the money going to come to pay for them?'

A few days later, her spirits had still not lifted. 'I am beginning to feel uneasy about the success so called,' she wrote again that evening. 'Already so many people coming, so much work to be done – the pressures on us all getting heavier – will it spoil the pleasure we have had in the past? It is reassuring financially but I feel the atmosphere is changing. Perhaps it is when I am tired and depressed that I feel that way. For me it has been such an intimate garden shared with garden-minded public – do I sense a different lot coming – I am not sure.'

She did not have long to think about such things since the stream of invitations continued to arrive. Lambeth Palace called. Would she come and meet the Queen there in two weeks' time? No, she sent back her reply. She was recording BBC Radio's *Gardeners' Question Time* at Barnsley, home to another gardening 'queen', Rosemary Verey. They called again to try and persuade her to change her mind but Beth was adamant, saying that although she was honoured, she would rather spend time with 'the young generation!!!'

A daunting invitation that Beth did accept was to appear on BBC Radio 4's Monday morning talk programme, *Start the Week*. *STW* has always been known for its cut-and-thrust discussion among intellectual glitterati and Beth was understandably nervous of her well-known co-contributors that week: outspoken feminist academic Germaine Greer, egghead quizmaster Bamber Gascoigne, *Beyond the Fringe* wit and opera buff Jonathan Miller and arts commentator Melvyn Bragg. Greer she knew already as she had been buying plants from Beth since Chelsea for her home in Great Chesterford, Essex, close to where Beth had spent her happy early childhood. 'Germaine is very enthusiastic – picking up plant names fast, but not yet certain what to do with them,' she wrote on the evening before the live broadcast. But she admitted to feeling intimidated by, in her words, 'that scintillating line-up'. The night before, she collected some 'dramatic' plant material from the garden to take with her and went to bed wondering if any of the men knew anything about plants.

The following morning, a taxi sent by the BBC arrived at 6.15 a.m. The driver had to wait a few minutes while Beth finished packing her plants. Among the flowers and foliage she took with her to Broadcasting House were *Trillium sessile, Uvularia perfoliata*, a selection of fritillaries including *Fritillaria pontica, F. pyrenaica, F. acmopetala* and *F. verticillata* as well as *Rheum palmatum* with its leaves of a rich beetroot red, *Euphorbia robbiae, E. wulfenii, E. griffithii* 'Dixter' – and flowers of *Gladiolus tristis* cut from the tunnel. They had all been packed the previous night into a large dustbin liner. Beth found them 'fresh as paint' the next morning. The car sped down the empty roads and Beth was the first to arrive, with Germaine Greer soon following. After the men appeared, Beth said she suddenly had 'a fit of feebleness', and again wondered how much any of them, except Germaine, would know about her subject. 'Would they be able to chip in or challenge me?'

Later that day, Beth reflected on the occasion. Germaine Greer she found very easy and informal – 'I like her. Very, very quick brain with an easy command of English.' The three men Beth thought 'all a little on stage being themselves, except Melvyn Bragg who acts the part of this ordinary fellow'. Her conclusion was that, in the end, it all went well enough, though, she felt, '50 minutes is quite a long time to keep a ball in the air.' What they all made of Beth's large assortment of plants is not recorded. Afterwards there was champagne and orange juice. She had a Buck's Fizz. Andrew, she noted later, was quite surprisingly pleased with her efforts. 'Most flattering – I had not impressed myself.'

WITH A HOLIDAY to Scotland with Christo confirmed for the middle two weeks in June, Beth had to work hard to catch up with jobs in the garden. In particular, there were young plants in the nursery that needed attention. These were a new stock of asters, heleniums and others that she had brought back from her visit to Germany with the Countess and then Holland with Romke van de Kaa.

The drama of the Great Storm the previous autumn somewhat overshadowed a smaller drama that had played out during Beth's stay with van de Kaa. The day after he had driven her from Laufen to his home in Dieren, he suggested a visit to the nursery of Ernst Pagels just over the border in West Germany. Pagels was a renowned nurseryman, especially for his development of garden grasses and species perennials. Beth was anxious to meet such an important figure in European horticulture. The pair drove for three hours towards the German border and, as they approached it, Romke reminded Beth to get her passport ready for inspection. Beth's face drained of colour. In her excitement that morning, she had carefully left all her valuables, tickets, cheques and passport on her bedside table. Romke later remembered her feigned innocence: she quietly asked whether they really needed passports, to which his answer was yes. Undaunted by the burly border guards who refused to let them cross without Beth's papers, Romke carried on up winding narrow roads through woods and small villages until he found an unmanned crossing. With enormous sighs of relief, they reached Pagels's nursery in Leer, East Frisia. It had no name at its entrance, just a small, modest drive that led to the stock beds. These were all covered with crushed bark, as were the paths.

Beth immediately felt at home there. 'We found a nursery after my own heart,' she wrote in her diary. 'A very human atmosphere – certainly not a factory.' She was impressed with the long piles of compost, seeing that even the woody stuff was not burned, just left to rot slowly with roses and hops growing over them, making a home for hedgehogs, frogs and toads. Everywhere there were jars of beer, set among the beds and pots to catch slugs and snails. It was, she concluded, an organic nursery but obviously commercially successful, enough to allow Pagels to observe and experiment.

Eating cheese-filled buns to refuel after the long drive, Beth and Romke walked around in the rain examining the plants. They looked especially at the various miscanthus grasses, many in flower. Pagels had initially grown them under glass, collecting precious seed and refining until he produced the best forms. Some were tall with upright flowers or arching, others were tiny, and some had good autumn colour. Finally, the seventy-four-year-old Pagels appeared and Beth was thrilled to be told he had a copy of her catalogue. 'Although we could not communicate in words,' Beth later remembered, 'we did in feeling and knowledge of plants – but so much I could learn from him I'm sure.' They left shortly after, the car full of plants that they had bought and paid for and delighted with the many more specialities that Ernst Pagels insisted they take with them as gifts.

Their elation did not last long. This time, Romke's secret border crossing was manned, and his somewhat unwise attempt to 'run' it resulted in his car scraping the legs of a particularly dour-faced German guard. The renegade couple were arrested. Charges included entering Germany without a passport, no driving or motor insurance papers and, most worrying for Beth, no plant export certification. Held at the local station by six huge men and a hefty blonde policewoman, all with guns, Beth also, perhaps unwisely, complained about the bureaucracy and form-filling. Would you rather go to court, she was asked, to which she meekly replied no. The couple were only too conscious of the fact that Beth was due to give a talk that evening back in Romke's home town in Holland.

Eventually, calls were made to Romke's local police station to which his wife, Adriana, had to take Beth's passport for validation. Fines were paid and, most importantly, the two were allowed to leave with the precious plants. A stop for petrol and two more cheese buns later, they got back with two minutes to spare before the guests arrived. Beth dashed upstairs to change to re-emerge, as she wrote, 'as the famous lady-gardener!'

The visit finished with a trip to the nursery of Piet Oudolf, who had been Romke's business partner before a split. Beth knew Oudolf, his wife, Anja, and their family since they made regular visits to Essex whenever they travelled to the UK. She was also already aware that the Dutch influence on plants was being underestimated in Britain. 'We tend to think we are the only ones who have unusual plants,' she told her hosts, 'but the Dutch are coming along fast with species plants.' One guest replied that while they might know how to grow plants, they would never be able to make a garden like Beth Chatto. Once again, Beth signed off her diary that night with her favourite phrase: '"Well, there now," as Cedric would have said.'

TRAVEL NOTES 3

A trip to Scotland with Christopher Lloyd (June 1988)

IN JUNE 1988, Beth and Christopher Lloyd went to Scotland for a rare holiday together. The arrangements were made by Christo, ever the social butterfly, who seemed to have connections everywhere. Beth was excited by the itinerary, which started with them catching the night train to Aberdeen, where they picked up their hire car. These edited highlights follow the pair as they zigzagged across moorlands and mountains with Christo driving and Beth navigating, visiting gardens and nurseries and staying in varying degrees of comfort with some of Christo's many friends.

Wednesday 8 June 1988

It was lovely to see Christo again. We sat in his cabin and drank whisky and water and ate his Dundee cake. All very reassuring. Drove to Crathes Castle – met by head gardener. Interesting but not a good garden, misses the hand and mind of the owner/gardener. The castle itself very attractive inside – cosy smallish rooms attractively furnished. Drove more than two hours to Jim Sutherland's* alpine nursery [Ardfearn, near Inverness], only begun two years ago but full of treasures and well laid out.

Thursday 9 June 1988

Marvellous warm sunny day. Walked round Alan Roger's* garden [Dundonnel, near Little Loch Broom] with Christo. Lots of good things, fine trees and shrubs – but now too overgrown. Very sad – seems his gardener is not good – not up to the job, yet he cannot dismiss him. Such a shame to let protection of employees lead to such neglect of common duties. But I do understand how difficult such situations can become.

Friday 10 June 1988

Hot and sunny all day. Incredible weather, clear, sunny and very warm. Seems to be the opposite in the south – wet and the coldest June day for x years!

After breakfast, Alan drove Christo and I almost down to the end of the loch where it meets the sea, to the Ardessie Falls where we began a seven-and-a-half-hour day of walking and scrambling. We were to climb up behind An Teallach. It made good walking but we saw few flowers. Walking up beside the magnificent waterfalls, lots of ferns, sphagnum moss, rowans and twisted little birches. They are eaten off where the sheep can reach them. Christo carried binoculars, we both carried sandwiches in our pockets. I found going down almost more difficult than going up. But although tired, legs aching a bit, I felt pleased and a little surprised I had made such a walk with no practice except rushing about the nursery.

Saturday 11 June 1988

Another very warm sunny day. I tucked two packets of tomato, lettuce and cheese sandwiches packed for us by Kathy into the lower leg pockets of my drill trousers and we drove partway along the peninsula which we had seen opposite us yesterday. A magnificent day for such a beautiful walk. The loch lying to our left, deep jade green over pale rocks immediately below us fading into dark blue and finally steely blue in the distance. As we walked on and on, rounding buffs and headlands, the loch opened out to meet the sea. During the day we found four orchids – butterfly (pale cream), spotted orchid, fragrant orchid – very deep purple and slim narrow heads heavenly perfumed, and the marsh orchid. Lots of ferns. After 14 miles flopped on my bed for few minutes before lots of tea and coffee cake. Picked heather out of my socks.

Sunday 12 June 1988

Another scorching day – After breakfast, Christo and I sat outside the front door and roasted ourselves in a perfectly still very warm morning. Air clear as glass above the mountain tops. Alan appeared and we had a conducted tour through his arboretum. Then we drove to Inverewe to have lunch in the house with the new curator, Professor Douglas Henderson* and his wife, Margaret. He was Regius Keeper at the Royal Botanic Garden, Edinburgh.

Monday 13 June 1988

Last evening mist rolled across the face of the mountains like steam from a kettle. Woke this morning to an overcast sky. Had breakfast sitting by the window all dressed and ready to go.

Found Christo irritably eating his alone and feeling uncomfortable with his breakfast on a tray in the sitting room (dining tables much more to his liking).

Drove to Fort William down Loch Ness – a beautiful drive for three hours until we reached Oban where we were to meet the ferry. Just before we stopped at a garden centre, the biggest in the North West it is said. Probably was – huge selection of plants, trees and shrubs but no atmosphere – no organisation – just heaps of plants, many in need of water. No personal touch – depressing I thought to work there. Later was told that everything was bought in from Holland! We met John and Hilary Carr, Christopher's ex-stockbroker and wife. Charming friendly people. Over lunch in the hotel near the ferry, I learnt that John knew Roy Carr (David Scott's cousin) very well. 2.30 p.m. went by open ferry across to Isle of Lismore. Very attractive home, beautifully furnished and comfortable, both so kind and hospitable – all very relaxing – after looking at the garden Christo and I both went to our respective beds – I slept till 5.45 p.m. – comes of drinking half a bottle of wine for lunch. Decided no more alcohol today. Just as well. John and Christo consumed an alarming amount at dinner. I fear for my liver much as I like wine.

Tuesday 14 June 1988

Woke to a total fog after yesterday's warmth and sunshine. Hilary and John had prepared a picnic lunch and off we set all in anoraks, climbing boots and equipped with walking sticks to ride in the Land Rover to the end of the island about five or six miles away – a very narrow winding road – just a tumbledown cottage here and there. But lovely plants, great patches of iris and further along a patch-like haze of ragged robin – sheep and lambs everywhere. Lots of wild orchids – Christo and I got out and botanised, finding all kinds of interesting things.

We had a delicious lunch under a bluff of rock sheltered from the wet clouds that still swept over us and hid from view the islands that we should have been seeing like Mull and others. From time to time the mist cleared and we could see what we had been brought to see but I was not too bothered – the weather seemed right for mountains – too much clear sky and hot sunshine takes away some of the mystery of mountains and my ideas and imagination of Scotland and the islands. Besides it concentrates the mind and eye on the immediate scene. After lunch the Carrs went back to the Land Rover and Christo and I explored the plants of the cliffs, finding lots of interest in the plants. Gradually the mist cleared and we could see the islands, rocks and seaweeds – the coastline of rocks. Stopped on drive home to sit in a little alp of orchids, grasses – all so pretty.

Wednesday 15 June 1988

John and Hilary drove us to the jetty to catch the ferry to the mainland. It was an overcast and drizzly morning. Felt quite sad to say goodbye. Hilary repeated the invitation to come again with Christo or alone – very kind of her. We collected our plants from the garage. We set off along Loch Ness, then Glen Coe, Rannoch Moor – by small roads with wonderful scenery all morning. Stopped at place making bone objects, bought combs for Diana and Mary, small bone spoons for myself. Christo also bought spoons and salad servers.

Arrived at home of Kulgin Duval and Colin Hamilton [Frenich, in Perthshire] midday, wonderful place built from U-shaped range of farm buildings built of stone and granite. They have good taste and enough money to buy beautiful things. Spent afternoon in garden – many plants I haven't seen before. Sat and talked and almost dozed in their meadow with beautiful views of Mt Meadows. Colin and Kulgin seem to know everyone. Their professions – bookseller and publisher – have led them into the world of celebrities in the art world and a wide circle socially. But although they may seem to have the best of everything, they are so very nice, so generous, so full of enthusiasm and have such good taste. I find them totally likeable and inspiring.

Thursday 16 June 1988

Spent afternoon with Lady Ruth Crawford* and eventually her husband, Lord Robin Crawford, ex-MP for St Albans. Fascinating garden – full of good plants – an exuberant garden, like its maker, Lady Ruth. She was charming, enthusiastic, warm and friendly. Her gardener, Donald Lamb, was with us too. Both must work incredibly hard. Happy conversations. Christo teases me a lot but I tease him back. We are having a wonderful holiday – I am loving being here.

Friday 17 June 1988

Again the day began shrouded in clouds draped across the mountains. After breakfast we went to Cluniemore, [Pitlochry] to see the gardens of Mrs [actually Lady] Myra Butter.* Her husband owns most of the mountains around the house and garden while she came from a wealthy family at Luton Hoo (sounds familiar but I do not know why). Appears she is a distant relative of the Tsar but we did not have the pleasure of meeting her since she and her husband are at Ascot this week. Seeing her garden, I am pleased since it would not have been easy to enthuse about it. Not a patch on Lady Ruth's exuberant planting we saw yesterday. Spent rest of the afternoon in my room writing letters. I think that the walls and curtains here are made of Fortuny fabric. Looks like silk, beautiful soft texture and colours in single tone – yellow rose or green on cream background, floral and leaf patterns, made in Venice (or Vienna I forget – Venice?!). Colin showed me his copies of *Curtis's Flora* [more probably Robert Thornton's *Temple of Flora*] valued at around £20,000 – hand painted over printed drawings. Beautiful colour. Also special book bindings they had commissioned artists to do.

Saturday 18 June 1988

Woke at 6.30am, overcast but much colder than at Alan Roger's place on the West Coast.

Glad to put on my white ribbed pullover over jade green trousers and shirt. But already at 7.30 a.m. thin sunlight is moving across the tree lined grass fields on the slope across the loch. Our last breakfast at Foss, oatmeal porridge, oat cakes and honey. I had put all my luggage by the car, including all the boxes of plants we have carried round Scotland. I hate saying goodbyes but this I felt the most sad about. Colin and Kulgin have been the kindest, most thoughtful of hosts. As he embraced me Colin gave me a little parcel from him and Kulgin. This evening, in another bedroom, I find it is a bottle of Floris bath essence of limes. They have already given so much – imagine giving me a leaving present. It is I who am indebted to them. We set off once more – a beautiful drive through leafy lanes with precipitous ravines alongside. Here we stopped and Christo looked for and found the hard fern, halfway down a sheer precipice in a wood.

Made me dizzy to hang on to a tree and peer down to look at it.

Then after a drive of about three-quarters of an hour we found his late cousin's garden in Keillour Castle, a Victorian mansion complete with turrets. The gardener, Alan, showed us round and [I] became more and more impressed and astounded as the garden unfolded, packed with good plants, placed so well – allowed to feed themselves yet kept within bounds – producing a natural and most harmonious effect. The cousin, Mary Knox Finlay, and her husband, Major W. C .Knox Finlay,* both of whom received the VMH, must have been remarkable people, creating this garden with TWO gorges, one over 50 feet deep, steep sided, planted with mature trees, species rhodos, primulas, *Lysichiton* [skunk cabbage] and many other suitable plants. It looked as I imagined a Himalayan mountainside might look with a few exotics added.

Sunday 19 June 1988

Drove to the garden of David and Melanie Landale,* Dalswinton. They know little or nothing about gardening and spend all week in London. Have a HUGE Victorian pile of a house, 18,000 acres of land and plenty of fine trees, rhodos, a tennis court, a newly planked laburnum arch – too long and narrow – too narrow in particular – Christo didn't hesitate to say so – after being so thrilled with the beautifully made iron framework I could feel Melanie's dismay as if it were my own. She is a sweet person, genuinely wanting to learn how to cope with her garden but no real chance when she is away so much.

Next, we went to Threave, another huge house with 60 acres, converted about 30 years ago into a school of horticulture and the land developed as a very interesting botanic garden. Planting not artistic but wide range and well kept. Two men, Bill Hean,* the administrator, and Magnus [Ramsay], a tall sunburnt botanist, in charge of the students showed us round. They were pleasant but a bit stiff.

Couldn't see half, of course, but was impressed by the scope and scale of planting achieved in 30 years. In the vegetable garden saw gooseberries trained as cordons – we must do that. Several plants new to me I would like, made notes of them. Shall write but not certain these two rather dour souls will remember me!

Monday 20 June 1988
Drive south to Northumberland to home of Bob and Rosemary Seeley.*

We came to signs for Gretna Green. 'No,' said Christo to me, the navigator, 'we don't want that.' 'Why, are you nervous?' I asked. I think we both enjoyed feeling relaxed enough to tease each other from a position of safety.

The drive all the way down to the Seeleys' was a pleasure. Countryside became rolling rather than mountainous – those we had left behind in the highlands. Now rich meadows, stone walls, roadsides rich with wild flowers – marguerites, buttercups, small daisies, birds, many dainty grasses all edging the roadside in profusion – wild rose, honeysuckle. Saw a slope pink with ragged robin mixed with a nice dark purple thistle. Then in damper soil, masses of *Geum rivale*, *Dryopteris pseudo-mas* and the broad buckler fern. I have loved the ferns up in the north, everywhere – bottoms of walls, hedges, banks. We stopped and explored a limestone escarpment for plants.

Tuesday 21 June 1988
An overcast day but no rain. This part of the country is in a state of official drought. It looks green and beautiful, but a rain would do good.

After breakfast, we put on anoraks, wellies and umbrellas in the car just in case. Bob drove us to the house and garden of Dennis Davidson, a retired storekeeper he called himself. An amazing man, tall, thin as a rake, he (and his late wife) has very good taste (and the necessary money) to have made a very beautiful home and garden. A large rambling stone house dating from the 17th century with additions of course. When we stepped out a chaffinch appeared at our feet – I bent to look wondering if something was wrong but it sat just a foot away from me and cheeped. Dennis appeared and welcomed us and fed the bird from his hand. All morning the bird and the black tit [probably a coal tit] kept with us in the garden and took peanuts from my hand. Christo thought I wouldn't like the garden, certainly there are too many conifers, too much of many things in the small space, a busy garden. Overall, I found it attractive and astonishing, since all the work is done by Dennis.

Then we drove about half an hour to Belsay Hall – another astonishing and different place. . . . I found the building most depressingly ugly . . . but the gardens were another matter.

Thursday 23 June 1988

Leaving Bob and Rosemary, Christo managed to lock the front wheel of his car with the back mudguard of Bob's – the park place was a bit confined – so we had a sticky few moments before we were extricated. I hate goodbyes and everyone watching when I am driving – so I felt very sorry for Christo. However we were away, calmed down and had an easy drive down to Rutland to have lunch with John Codrington. I was navigating and as we drove part way round a great estate I was suddenly amazed to find we were outside the park gates that told me I am close by Geoff Hamilton's place and there was the red posting box and little hole in the hedge which leads to his house and the BBC television garden! Such a surprise – John's garden is almost opposite on the other side of the reservoir. Charming old stone cottage with mullioned windows, roses over everything, a border of ox-eye daisies and pink campions leading to the front door. Inside a low sitting room, flower arrangements on low tables fresh and dried. Paintings, birds' eggs, books all kinds of objects collected from his travels – all carefully arranged – lots of low light. Much more light from the garden side where he had expanded the room into a conservatory. There, on the wall, was a huge mirror picked up cheaply because it was too big for the shop.

The garden is a child's paradise – wild overgrown and yet not entirely, his underlying design had it safe from total chaos and much of it is quite beautiful. It began as a flat open site without a tree – now it is a series of secret gardens. The day was warm and sunny, Christo almost fell asleep in his chair, he nodded and swayed so I said we must leave (it was 3 p.m.). We drove a few yards, then Christo fell asleep in the warm grass by the reservoir.

11
Honours and opera

INVIGORATED AND RELAXED after her holiday in Scotland with Lloyd, on her return to Elmstead Market, Beth wasted no time in checking the changes in the garden over the two weeks. 'I am glad I have made the decision not to do Chelsea but to concentrate my efforts here,' she wrote in her diary that night, 25 June 1988. 'There is plenty to be done [but] it looks very exciting now.'

Two days later she and her twin brother Seley celebrated their sixty-fifth birthday. Having first picked elderflower blossoms to make muscat-flavoured syrup, she wandered into the nursery where all the staff called out 'Happy Birthday'. Later, after going through her post and opening some presents in the office, she was called through to the staff room. There she found them gathered. They presented her with a parcel and a card that everyone had signed. Inside the parcel was a framed pressed flower picture of Cedric's narcissus, snowdrops and hellebores. Later that evening, Beth wrote how she had enjoyed the whole day, made special by her family and staff.

Beth's social whirl continued. The *London Evening Standard* wanted to feature her again and asked to send a photographer in the next couple of hours. Beth agreed and admitted to quite enjoying the experience for a change, since the photographer used a long lens and she did not feel as exposed as usual. This was followed by an invitation from Christo to go to Glyndebourne in August – and could she make the picnic? It would be five men and Beth. 'What a fright,' she thought, but since it was to be *La Traviata*, she relented.

But before that, she had what she later called one of the most memorable days of her life. On Thursday 14 July 1988, she went the short distance down the road from Elmstead Market to 'our university', Essex, to receive an Honorary Doctorate. Accompanied by her brother Seley, she first attended a reception at the Wivenhoe

Beth receiving an Honorary Doctorate from the University of Essex, June 1988.

Hall (now Wivenhoe House) hotel, an attractive red-brick period building which sits in the middle of the 1960s campus with its brutalist modern architecture. Beth, dressed in a biscuit 1920s-look longish linen skirt, short jacket and striped silk scarf, was relieved to see lots of people she knew from the Colchester Officers' Club, and other regular visitors to the garden. At lunch she sat on the top table with Sir Robin Day, the formidable television political interviewer, and former diplomat Sir Andrew Stark, Pro-Chancellor of the University, who were also receiving honorary degrees. Beth enjoyed talking with them and, somewhat to her surprise, found the bow-tied Robin Day, notorious for his penetrating television interviews, easy to talk to, having feared he might steamroller her. Then, robed in scarlet with a black flat cushion hat (which was too tight – but the larger size fell over her ears), they processed behind another robed figure carrying a silver mace into the Great Lecture Hall. A huge semicircle of faces disappearing up into the roof slowly came into view. Somewhere an organ rumbled. Beth later wrote that she felt ecstatic and astonished that it was her in this pageant-like recession – and a little scared.

Still basking in the excitement of the event, she then went down to Great Dixter and on to Glyndebourne to see *La Traviata* with Christo and his party. As she had promised, Beth spent the morning helping Christo with lunch and making last-minute preparations for the picnic. There was tongue cooked by Christo accompanied by Beth's *pan bagnat*, with green salad and freshly shelled 'Hurst Green Shaft' peas ('marvellously free of maggots and well-filled pods'), and flowers – pale yellow day lilies and nasturtiums – served on wooden boards and platters.

They drank champagne on the terrace before lunch, then dispersed to their rooms to 'don our finery', the men, as Glyndebourne tradition required, changing into dinner jackets, while Beth wore an Indian silk dress. It was not, she thought, particularly ravishing but she felt comfortable in it with its soft colours and clinging, feminine material. It was very different, she noticed, from some of the creations worn that evening, though some women were, she thought, very stylish and attractive.

It was a warm still evening and they all found the performance breathtakingly beautiful. 'I could not bear it to the end,' Beth wrote that night. 'How could I take any more – it was so emotional.' More prosaically, she was also delighted with her picnic, reporting that it was a great success with Geoffrey Gilbertson, the Glyndebourne front-of-house man, saying it was one of the most original he had had. 'A thrilling day for me – I loved every moment.'

The next morning was spent going round the Dixter garden with Christo, making notes of things she would like to take home. In a rare moment of restraint, she had planned not to take any plants, since she already had so many still to plant out. But the temptation was irresistible, and Christo was so generous, so anxious, Beth felt, to share this or that special plant (in this case ferns) with her. What a pleasure, she thought, to share the love of growing plants.

During a stay at Great Dixter, Beth sketched the flower arrangement placed in her room.

Beth was always grateful for Christo's hospitality at Dixter. She loved the place and felt relaxed and happy there, as if she had always known it. She loved its apparent changelessness, the size and grandeur of some of it yet also its simplicity and homeliness. It reminded her a little of Benton End, and Christo was opening up new worlds for her much as Cedric used to do – only, she thought, in some ways more so. She enjoyed his friends and felt they had become her friends too. 'It has', she wrote that evening, 'been the happiest of visits – good conversations with warm friendly and interesting people.' As always, their friendship was almost as

Beth and Christo's fascination with plants was the foundation of their enduring friendship.

much about food as it was plants. Before she left, she showed Lloyd how to use his new Magimix food processor. 'He was nervous of it at first,' she wrote, 'but I am sure will find it useful. He wants to make sorrel soup.'

Back home at Elmstead Market, the pressures of publicity continued. The actress Penelope Keith, best known for her television performances in *The Good Life* and *To the Manor Born*, was a keen gardener and came with her husband to make a programme in the garden for Thames Television. On arrival, they said they wanted to see the garden by themselves so it was not until later that Beth went out to meet them. They were soon on first name terms but they each asked permission and Penelope said how she hated it when people (strangers) used her name without even an introduction. The next day, the Thames film crew were in the garden from

8.30 a.m. till lunchtime, making shots to put into conversations Beth and Penelope had filmed together the day before. Beth was paid £50 – not a princely sum, she thought, and could not help wondering what they had paid Penelope.

Soon after, she was cheered by a visit from Piet Oudolf, who arrived on his way to catch the boat back to Holland. 'Such a charming man,' Beth commented. 'Good to look at, tall, blond, about mid-40s. But with such warm generous feelings too about sharing plants and our pleasure at possessing them.' Oudolf went round the nursery meeting various members of staff, saying how impressed he was and how clean everything was. He joined Andrew for a sherry before he went off for his regular evening pub visit when Beth and Piet had the empty nursery to themselves for Piet to pick out plants before leaving to catch his boat.

Beth also decided to drive up to Bressingham near Diss on the Suffolk/Norfolk border to see her old friend nurseryman Alan Bloom. Driving under a darkening sky which eventually turned to rain on the windscreen, she felt a foreboding that the visit wouldn't sparkle – it didn't. 'Poor Alan,' she later recalled, he did not immediately recognise me. He must be 83 coming 84 – still amazingly strong and active for that, but no social charm! Offered neither coffee nor lunch when I left after two hours and the long drive. But gradually he became enthusiastic to show me his plants – holding me back to see more when I was making a move to go. So unlike the old days when he greeted me with barrow, fork and labels. [This time] he did not offer a single cutting. I didn't need it but the change in attitude struck me.' The day ended with a visit from Hans. It was a chilly early autumn evening, and Beth lit the fire. They sat and talked for two hours. It was, Beth wrote in her diary later that evening, 'both full and frank' but she felt it was a relief to them both. It seemed that they had turned a corner. 'Now I think we can at last be old friends.'

TRAVEL NOTES 4

A trip to Yorkshire and Scotland (September 1988)

In September 1988, Beth returned to Scotland to speak in Edinburgh, taking the opportunity to visit friends in Yorkshire as well, including the RHS's garden in the north, Harlow Carr. While this was less of a holiday than her recent trip with Christo, these edited highlights show that, as always, gardens formed the staging posts of her journey north and south.

Friday 16 September 1988

Today I drove to Yorkshire to stay with Pippa and Arnold Rakusen who live near Leeds. Andrew rang at 7.30 p.m. to see if I had arrived! I was amazed and delighted. Was just about to ring him. He has never rung me before when I am away.

Saturday 17 September 1988

I thoroughly enjoyed walk round Harlow Carr. Pippa is a director of the gardens and has been for eight years and will have another six. She has put in a lot of good ideas, it all looked very orderly and well kept. Her views on colour harmonies and foliage groupings are well carried out.

Sunday 18 September 1988

Spent most of the morning in garden with Pippa – very interesting – very good foliage effect predominantly with trees and shrubs but also makes very effective use of small spaces, packing plants tightly together, no room for weeds.

At 6.30 p.m., Pippa went on stage and made a very graceful and generous introduction and the evening began and lasted one and a half hours. I showed 95 slides. Afterwards there was great enthusiasm. Many shook my hand and said it was marvellous . . . All my books and catalogues were sold and many brochures. I think I can say it was a success for them and such a relief for me.

Monday 19 September 1988

Today was a highlight – Pippa took me to see Jim Russell at Castle Howard – I felt sure I must have met him at Sunningdale with Graham [Stuart Thomas] when Andrew and I visited there long before we built the house at Elmstead. On first sight he was tall, portly and cautious – or perhaps reserved – but as the morning went by he became more enthusiastic until by the end of the day he was positively glowing. I felt he began feeling it might be a bore to take two strange women round (although he had met Pippa there before) but between us we all enjoyed his work so much and were so appreciative of both his talent and the plants – it was a mutual pleasure for us all.

We spent the morning in a vast woodland garden with paths of springy moss meandering up and down the sloping terrain between rhodos, ferns, hydrangeas and many more appropriate shade-loving trees, shrubs and plants. He travels extensively especially to Japan and is much influenced by seeing plants in the wild – so his plantings and the self-sowed seedlings that result appear natural even though many of them are exotic. I found myself writing notes busily to retain some ideas to help me plant my little scrap of woodland when Gerard and John have cleared up the mess from the hurricane. We returned to his house for lunch with his sister – in the old dairy – converted and looking like an 18th century house. After lunch back to the woods and to see the beautiful mausoleum and a bridge spanning a great lake. The Howard family still own all the land to the horizon – beautifully patchworked with farmed fields and woodland. Finally, Jim drove us in his car around another new enterprise – his arboretum – an area newly planted (less than 10 years) of thousands of rare trees, many species of oak, willow, chestnut, poplar, beech etc. Quite an amazing man – and an amazing day. It was gone 7 p.m. before we left – home by 8.15 p.m.

Wednesday 21 September 1988

A great day at the Edinburgh Botanic Garden. Beautiful lecture theatre, seats tiered, perfect acoustics, no need for mike. Slides went through well – automatic adjusting of focus. Spent rest of morning till 12 p.m. wandering in garden (just before had to have photos taken in the water garden for press) – Garden immaculately kept. Almost too much so, not a weed, untrimmed edge other than newly shaved velvet lawns. But lots of good plants in less formal areas. Was entranced with the fern house and the cactus house where I found names of some of my succulents.

First among the audience was Lady Ruth Crawford with Donald her gardener. She immediately invited me to stay tomorrow – that will be very pleasant I know. There is so much I could and should see here in Edinburgh – but I will enjoy being with friend in Balcarres. People had come from far and wide, one lady 160 miles but said after how well worthwhile – the talk went well – a most receptive audience – lots of questions.

Thursday 22 September 1988

Arrived at 1 p.m. at Balcarres House – recognised the garden I saw in the summer along the approach drive. Found my way to the garden door – saw Ruth in the great kitchen who hurried to meet and embrace me. How reassuring to have such a warm welcome to this vast historic and beautiful home.

The rest of the day was bliss – we had lunch on the round table in the window, homemade soup, delicious mixed salads from the garden then looked at her drieds hanging from tall clothes airers and standing in jars. 'The Fairy' rose hanging above the Aga, even double peonies have been dried. I must learn from her and experiment more. Then spent a drizzly pm in the garden with Donald Lamb, her treasure of a gardener. They make such a team.

Friday 23 September 1988

I spent the morning in the garden (under umbrella) with Donald digging old varieties of border phlox and a superb form of pulmonaria. Ruth had started a cold so stayed in and did the flowers brought in from the garden by Donald looking magnificent with his curly yellow beard topped by a huge felt sombrero because of the rain.

It has been lovely to be alone with her, less intimidating for me than if all the family were there – husband and a married son with wife coming this weekend.

We had quite an intimate conversation about health – she has had two operations on her thyroid – hers swelled like mine – a kind of nervous breakdown also similar and problems with daughters (eldest) also similar – the difficulty of accepting each other as adult women – of having those emotional upsets before eventually we can talk and understand each other. I wonder if it is because she is European – Swiss – that she is so open and frank with me (not effusive). Few English great ladies would be so I think.

Left Balcarres about 3.15 p.m., Ruth and Donald seeing me off. Drove to Kirkcaldy then followed the sun to drive north to Kinross. Then Glenfarg where I was to stay with Maggie and Brian Lascelles.* They have a nice house in the middle of the village, surrounded by wooded hills and pastures enclosing the village.

Saturday 24 September 1988

We drove for lunch to Douglas Hutchinson, another friend of Colin and Kulgin. I loved his house. He is a bachelor with taste and enough money to have made a fascinating home on different levels. Garden well-kept and pretty tumbling brook, a walled garden with mixed herbaceous shrubs and fruit and veg. Beautiful drive home across the Moors. After tea glad for bed for an hour – then dinner with Roger Banks, knew Cedric, and two widows, both very agreeable.

Sunday 25 September 1988

First, we drove to Branklyn, a small mountainside garden originally a gem, neglected or abused for many years but now has a much more sympathetic and understanding head gardener, David Tattersfield.* On arrival I like the atmosphere – small narrow paths, many ericaceous plants, ferns, shade lovers and fine and unusual trees and shrubs. Many large trees had been felled in the bad days. David joined me and was charming and knowledgeable. Gave me seeds of several things. Would love to go again another season but saw many good ideas.

12
Autumn anthem

WHILE BETH DREADED WINTER, she loved autumn. For her it was an opportunity to spend all day planting and take time to look around her. 'Every leaf that is left is contributing to a glorious conflagration of colour, harmonies, shapes and textures. The birches are a deep rich gold, the oaks a lovely lighter shade, with remains of hostas making a shade of sharper lemon. I stand and stare and dream of what else I may plant in the future,' she wrote after a day in the garden on a sunny November day. 'The autumn colour ebbs and then surges again through these wonderful sunny days. Illuminate it like a rich tapestry. I find myself just wandering round and just enjoying the new combinations of colour and form that keep emerging. In a way it looks as well if not better, than it has all year.' She talked to Christo on the telephone about the delights of the Indian summer and the amazing lingering feast of autumn colour. 'I said it was so beautiful – the most exciting scenes in all the summer – I felt like singing an anthem all day.'

Autumn was a busy time in the nursery, with customers coming long distances to buy plants. One day, Beth helped a garden designer from West Sussex who wanted things in sixes, somewhat to Beth's puzzlement. She preferred to plant in odd number groupings. She was also a little taken aback when he said he had heard that Christopher Lloyd spent his time in the kitchen and she spent hers in the vegetable garden. 'Touché!' she noted later. 'I must produce more results in the rest of the garden this coming year.'

Although she had a strong team working in the gardens, with her international fame came a regular stream of requests from horticultural students to spend time working there. An extra pair of hands was always useful if they could stay the course. Beth tended to favour those from Germany or Japan, who she found were the hardest workers and keenest to learn. This October's intake was a little different. Thirty-two-year-old American Doug Hoerr was already a qualified landscape architect and had worked for relatives who had a landscape contracting business when he came to spend several weeks with Beth to learn about plants. Having also done a garden design course with John Brookes in the US, he felt he had reached a point in his career where he needed to learn more about plants. At the time, Blooms of Bressingham had become one of the best-known and successful perennial nurseries. Hoerr had written asking if he might get some experience working for them. However, he later told Beth he was not enthusiastic about time spent there, saying that he thought the management were too remote from the staff. Beth was disappointed to hear that he was expected to do constant weeding rather than being given the chance to gain more valuable experience doing other things.

While he was working at Blooms, he kept hearing about the Beth Chatto Gardens and started visiting regularly to see what the fuss was about. He was so impressed that he decided to approach Beth and ask if he could come and work for her instead. She agreed to see him but was initially reluctant, saying she was not sure whether he would fit in as he was 'so American'. This was a little surprising, since Beth had already had one American student, Jack Henning, whom she got on with, and they stayed in touch through letters and seed exchanges. Since one of her German students had just left, she agreed that Doug could come and live in the caravan on site near her vegetable garden. The contrast between that and the Chicago life he had left behind could not have been greater. She soon had him out in the garden, double-digging alongside the rest of the team. One morning he remembered her saying to him that she never felt bad about making young people work really hard for her since it was all for 'such a good cause'.

There were compensating highlights. Three weeks after he arrived at the gardens, it was arranged that Doug would accompany Beth for a weekend trip to Great Dixter. This was a rare honour for a visiting student and a sign of how easy Beth found him to get on with. It was a treat for her not to have to drive and she also found Doug an articulate and entertaining companion. The journey went quickly and the pair were greeted warmly by Lloyd and noisily by his dachshunds. As usual, they did not arrive empty-handed, bringing with them red chicory and a dish of raspberries picked by Beth from her vegetable garden before they set off.

A rare treat – the gardens at Dixter viewed from the roof in 1988 taken during a visit with Beth's American student Doug Hoerr.

Lloyd showed them to their rooms, with Beth this time in Christo's mother Daisy Lloyd's big four-poster bed with its dark-blue-lined canopy, surrounded by faded Jacobean design cotton curtains and yellow bedspread. Doug was billeted in the little bedroom with the blue quilted cover. Beth could see at once that he was entranced by the magic of Dixter. The rest of the day went by with the two visitors in something of a dream-like state. They enjoyed each other's conversation and the beautiful environment, with the low sunlight making the house look quite different that day, Beth remembered. That evening was spent drinking whisky in the solar in front of a log fire. Beth woke the next day with a slight hangover but it did not spoil the atmosphere. Doug was delighted to be with two such good friends, listening to their back-and-forth

Beth with Crocus and Christo with Tulipa, David Ward (left) and Bernard Trainor (centre), an Australian student who now runs a landscape designer business in California.

banter about plants and disagreements. A highlight was when the trio went up on to the roof of the house – a rare treat – and saw the gardens and orchard laid out before them.

The weekend was rounded off with a trip to the nearby town of Rye, where Doug and Beth wandered up and down the sloping cobbled lanes, with Beth proud to show off its seventeenth- and eighteenth-century buildings. They enjoyed themselves so much that Beth forgot to point out the turning on the way home and they ended up just over the Suffolk border in 'Constable country', having to weave their way home through country lanes.

Doug stayed for four months and took Beth to see John Brookes, with whom he was to spend the rest of his time in the UK. Brookes, who had known of Beth since her Chelsea days, now ran a design school at Denmans, West Sussex. In many ways, he and Beth came at gardening from opposite directions. While Beth was a supreme plantswoman, Brookes had written in his seminal 1969 book, *Room Outside*, that 'a garden is essentially a place for use by people . . . not a static picture created by plants.' Yet Brookes was also able to see Beth's contribution to design with her use of scale, shape and texture. He later credited Beth with moving horticultural design on from what he called 'the "Manor House" style of gardening', a generous statement considering he was also a key figure in this transformation.

While Brookes was regarded as something of a revolutionary in the gardening world at that time, Beth always enjoyed her connections with some of the 'old school' plantsmen who had been such keen supporters of her nursery in its early days. While she was stacking labels in the pack house in late October 1988, her secretary rang to remind her to get ready for John Codrington's ninetieth birthday party at the Garrick Club in the heart of London's West End. Beth had known Codrington ever since she began showing, and she remembered him always coming to her stand and seeming enchanted by it. She never forgot the time he had asked Christopher Lloyd if there was any chance of getting a pure white nerine that he had seen at Dixter. Christo told him he must make love to Beth Chatto (from whom he had had the plant) and he said he would.

The Garrick is known as the club favoured by those in theatre and the arts. Beth marvelled at the walls lined with portraits of famous actors. She did not know many people at the party but enjoyed the atmosphere, with champagne helping her to mix and chat to strangers. She was home by 9.30 p.m. She was far happier sorting through plants she had acquired during her summer trip to Scotland and seedlings from her escapade with Romke van de Kaa the previous autumn, relieved and exciting to find they have survived.

At the end of the year, she reflected that she had much to look forward to. She had been cheered in October by a visit from Piet Oudolf with his wife and two boys. She took them round the garden but felt too exhausted to ask them all in for tea. Oudolf asked if she would like to go to Holland next August and be a guest at a lunch on his open day, something she felt she might well do. In November, Christo had rung to say the Australians would like her to go with him next September. This was after she had initially turned it down because of Andrew. 'What excitement!' she noted.

As often at the end of the year, Beth was philosophical in her last diary entry of 1988, rounding up her thoughts on many personal and international issues and acknowledging that it had been an extraordinary year. She wrote about Andrew being desperately ill in early spring, when she feared she might lose him, but how, with the help of his doctor, oxygen cylinders and a night nurse for a week, he pulled through. Since then, he had slowly regained the weight and strength he had lost and had had a relatively good summer.

She was thankful that the nursery had progressed well and that personally she had had a year of excitement with the awards, lots of articles about her, the garden or her writing. She was relieved to have finished her new book, *The Green Tapestry*, and had also seen the *Notebook* published in September. *The Green Tapestry* had brought its own challenges. Unlike her previous three books, which were illustrated with black and white line drawings, Collins explained to Beth that the public now wanted 'lots of colour'. 'How I dislike that phrase,' she wrote, 'in the garden or in a book.'

Beth's antagonism towards 'picture books' was to last many years. But it never extended towards photographers. They were generally welcomed to the gardens even though their lighting requirements usually meant an early start. Beth got used to looking out of her bedroom window in her nightdress at 6.30 a.m. to see garden photographer Jerry Harpur setting up his tripod by the Damp Garden's ponds. Harpur had begun photographing the garden on a regular basis after his first visit in 1979, and he became one of Beth's favourite photographers. She had great trust in his work and rarely asked to see his pictures before they were sent out to magazines. She even acted as a peacemaker when Harpur fell out with Christo during a shoot in Dixter's Tropical Garden. leading to a lifelong friendship between the three of them.

By the end of 1988, she had had many good reviews for the *Notebook* and had accumulated an envelope full of letters of praise from people saying they had lingered over it, not wanting it to come to the end. 'What more could I wish for? I feel more than blessed.' Throughout her career, Beth kept all the notes and letters of praise sent to her stored in an ever-increasing number of box files. That New Year's Eve in 1988, she also thought deeply about what was going on away from 'my little world', considering both what she saw as the miracles and the desperate disasters around the globe: 'Mr Gorbachev (and his wife, Raisa) have done more than any other political leaders to give a ray of hope towards a world without the dread of atomic warfare. He seems to be a genuine person, has tremendous charisma, I have not heard a soul speak ill of him – even though some journalists keep up the idea of the Russian Bear just having sheathed his claws. Doubtless it is necessary to be wary – not to disarm ourselves, mentally or practically completely, but it would be wonderful to be able to get to know the Russian people, not as the enemy.'

She was not so complimentary about the United States: 'America went through all the absurd charade of electing a new president. It all seemed very undignified, full of razzamatazz and fostering new images to get votes – UGH. Then the disasters – terrible earthquake in Armenia, ghastly fire at King's Cross Station underground, air crash at Lockerbie, Scotland killing 270 and caused by a terrorist bomb, probably Iranian [It was eventually proved to be Libyan.] Out of the earthquake many tales of unbelievable heroism and endurance. A mother buried eight days with a daughter who survived sucking blood from mother's fingers.'

Beth's own year ended quietly and happily. With her typical urge for order and inability to fully relax until things were done, she spent the evening that New Year's Eve cleaning and tidying the office, sorted her mail and then, finally, did a week's worth of ironing.

13
The gardens grow

BETH HAD BEEN SO BUSY throughout 1988 she had barely had time to miss doing the Chelsea Flower Show. As she wrote in her *Notebook*, preparing for it involved her for six months and meant potting up roughly one thousand plants. The bigger shrubs needed 'loving care' to be 'in good heart for the next year, as well as each year'. While it was exciting to have a Gold Medal, she had then found it disheartening to come home to a garden in clear need of attention.

By the late 1980s, the main areas of the gardens – the Damp Garden around the ponds, the Mediterranean Dry Garden close to the back of the house, and the garden bordering Park Farm's reservoir – were all well established. The reservoir had been started by Andrew when his father bought him the farm in 1930. When the land was sold to Hans Pluygers, he had enlarged it to help irrigate his orchards.

After Badgers' Wood was devastated in the Great Storm and the badgers left, Andrew agreed that Beth could take this area over to develop as a woodland garden. At first, he had been reluctant. This was one of the two parcels of wasteland which he had always deliberately left for wildlife. Beth had telephoned him during her visit to Scotland in September 1988, having been inspired by the woodland garden at Castle Howard in Yorkshire. Andrew's immediate response was a firm no. She could not make a woodland garden in Badgers' Wood. Eventually, as usual, he gave in. The area that was to be cleared was full of self-sown trees – a mixture of oak, ash and birch together with an understorey of elder and blackthorn – all native English species. A start was made over the winter of 1987–8, with Keith Page's son Gerard, who had now joined the team, in charge of clearing the damaged trees. Beth later told the story of creating this area in 2002 in *Beth Chatto's Woodland Garden*.

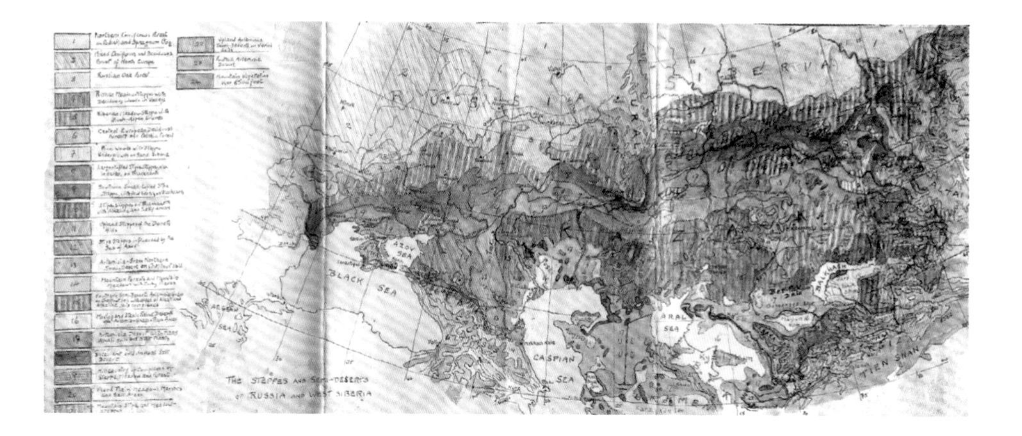

LEFT Beth and her team began work on preparing the woodland garden after the destruction left by the Great Storm of 1987.
ABOVE A hand-coloured geological map of Russia by Andrew Chatto, now part of his botanical archive.

But her bouts of depression were not diminishing. While her career and the garden were flourishing, her private life remained complicated. She had been married to Andrew for over forty years. Now the fourteen-year age gap was becoming more noticeable as Andrew aged and became more infirm. A pipe smoker until he gave up in 1986, he was affected at times with emphysema and every chest infection was potentially life-threatening. While Beth dealt with the public, Andrew still retreated all day to his study in the roof, his 'eyrie', where he laboured for hours over his research into plant ecologies across the world. In the days before computers and web access, he did this entirely from maps, books and encyclopaedias.

His only relaxation continued to be a nightly visit to the local pub, with Beth usually dropping him off and picking him up a couple of hours later. For the most part this mystified Beth, since it seemed to contradict Andrew's naturally solitary personality. Yet he was sociable in his own way and enjoyed chatting to people in the pub and 'putting the world to rights', making good friends of all ages among

Beth and Andrew celebrating their Golden Wedding on 7 August 1993.

the regulars. Just occasionally, Beth would join him with surprising success – as she remembered about a visit to the King's Arms in Frating just up the road from the gardens. 'We went, drank beer together and talked together as if we hadn't seen each other for a week!' Given Beth's busy life, in many ways that was true. 'We see little of each other all day,' she wrote, 'but when we are together we are never bored. Somehow we always manage to entertain each other in the evening conversations.'

As with many long marriages, there was a regular daily pattern to their life, a rhythm that when undisturbed can continue for decades. The disturbances, when they happened, always involved Beth. Andrew hated her going away and discouraged her from accepting long-distance invitations – usually just with a pained expression in his dark eyes, but occasionally more directly. It was the same when any new opportunity or challenge came up. In early 1994 she was asked to write about the Chelsea Flower Show for a German magazine, *Architektur und Wohnen*. 'You can't do that,' exclaimed Andrew, 'it's too long since you were there and you don't know what is happening now.' After so many years, Beth was getting used to such comments and went ahead regardless, knowing they wanted a personal account rather than a list of facts, later writing in her diary that 'the thrills and traumas of being an exhibitor couldn't have changed very much.' While Andrew's reason was always that he didn't want Beth to be upset by rejection, his sometimes challenging remarks indicate there was a deeper message of anxiety that he might lose her, despite her solid refusal ever to consider leaving him. On one occasion when Beth had upset their daughter Mary, Andrew tried to reassure Mary by pointing out that Beth was rather like a child who needed caring for and understanding.

In many ways, it suited them both that Andrew spent most of the day researching plant habitats in what Beth called 'his ivory tower', although she was kinder in her letters to Christo, calling it 'his little room' or 'little hideaway'. Andrew was approaching eighty, and Beth was in her sixties but with the energy of someone ten years younger. Increasingly, she was having to nurse him, on top of her many other commitments. Hans was still just across the field, which made it hard for Beth to distance herself from him completely. Although her relationship with him was no longer physical, having finished after her discovery of his other infidelities, his visits were usually a welcome distraction.

This small, ebullient Dutchman had been part of both their lives for decades. He would remain so, particularly since he had bought some of the land surrounding the gardens, known in the family as 'Chatto land', from them. They could never have imagined that some of that land, sold to Hans before the success of the gardens, might become useful for expansion, or, more importantly, necessary for access. For several years in the 1980s and 1990s, Hans remained a key figure in both their lives, with staff issues and rights of way a constant frustration. After his wife died, in 1985, he would almost daily make the walk across the fields to White Barn House, hunting Beth out in the greenhouse or gardens and inviting himself in for a cup of tea and a chat about the ups and downs of the fruit business.

As the popularity of the gardens and nursery grew, there were regular issues to do with land rights – and not just those involving Hans. When Andrew's farmland had been sold in parcels over the years, they were left with an area of 3 acres.

By 1990 they had been able to buy back around 12 acres in various chunks. But running between the garden and Hans's field was a ditch filled with blackthorn and brambles at the top of which grew the 300-year-old oak tree that dominated the view from the house. With the ponds made and the land on their side drained, Beth planned to clear the ditch of its brambles and bracken, having checked first with the tenant farmer. The ditch sloped down from the farmland and could not be seen, but was the source of the spring that fed the ponds. Shortly afterwards, an official from Essex County Council arrived to explain to Beth that the ditch and bank were council property. The issue was eventually resolved but not before Beth understandably became extremely stressed over the implications of losing control of the spring.

As was her habit now, she went to Dixter to get away from the problems at home. While she was staying with Christo in June 1990, he took her for a picnic at the desolate area of Dungeness on the Kent coast, just a few miles from Dixter – a bleak shingle bank bordering the eponymous nuclear power station. Walking back up the beach, they came across a black-painted cabin with yellow-framed windows and beside it some santolina. It was, wrote Beth to a friend later, 'as fine a plant of santolina as you'd never expect to see in such an arid place, like a great wind-blown fluffy ball come to rest in the shingle together with curious family groups of weird stones, bits of flotsam and jetsam – of wire and other things . . . Scarcely aware, we were led by curiosity round the back of this cottage where we were astonished to see the range of plants, all flourishing among sand and pebbles.'

They had happened on the seaside home of the film director Derek Jarman, who invited them in, answered their questions and took them into the tiny front room filled with more trophies salvaged from the beach, all grouped purposefully, like the garden. That night, Beth wrote in her diary: 'As we left I knew I had been very affected by the way his plants looked and flourished in that most unlikely of situations. We all paint very different canvases, but I had been encouraged by that brief glimpse of his palette, to make a gravel garden [at Elmstead].' Jarman himself was equally impressed by his visitors, noting in his diary that date that 'amongst the visitors to the garden today were Beth Chatto and Christopher Lloyd. They were taking notes and photos. I realised quite quickly that I was in the hands of experts. Beth knew the Latin names of every plant, and when she told me who she was I nearly fell off the Ness.'

Beth was equally amazed that Derek Jarman should have known anything about her, and was delighted when he later sent her a list of a year's tally of plants growing within a mile's radius of Prospect Cottage: sea kale, red and white valerian, sea campion, woody nightshade, plantains, bird's-foot trefoil, star of Bethlehem (this had run wild around an old wartime bunker, Jarman noted), creeping buttercup, scarlet pimpernel, cinquefoil, hound's tongue, yellow rocket (winter-cress), ivy-leaved toadflax, hop trefoil, Nottingham catchfly, and many others. (A couple of years later, after a visit to Beth's nursery, Jarman also sent her some seed of a mauve Oriental poppy that Beth had spotted on the Ness.)

Beth was excited by the list and wrote back bursting with enthusiasm for her new project. She wrote to Jarman: 'I am planning to make a Gravel Garden in front of my house, on an area of land, perhaps half an acre, where last year visitors parked their cars. It was originally grass, made grotty by droughts and cars – so not suitable for grass again (boring anyway). We have ploughed and subsoiled (a wonderful tool breaks up the compacted subsoil) and must add lots of compost now because we won't have the chance again. Then I look forward to planting it, inspired in no small way by what you have achieved at Prospect Cottage. I was so impressed with the robustness of your plants and the attractive contrast with pebbles. God knows yet what the final effect will be – I work it out as I go along, having first made lists of plants I'm pretty certain will stand the conditions, i.e., poor soil, very free-draining gravel, and low rainfall. There will be no irrigation. First the idea is to please and teach myself, but secondly, perhaps, to help my visitors to learn what they might grow where hosepipe watering is banned.' Beth's plant lists were often written on rectangular strips of corrugated cardboard cut from brown boxes, so much more practical to write on and refer to than paper, she thought, when out in the garden.

The inspiration for the Gravel Garden came to Beth from several places. In the introduction to *Beth Chatto's Gravel Garden* (2000), she credited a dried-up river bed she saw in New Zealand during her 1989 round-the-world trip with Christo. She also mentions Derek Jarman's Dungeness garden. However, she later put a sticker on the small red Silvine exercise book she had used for a trip to Ireland with Christo in June 1991 saying, 'In this note book I wrote how and where I found the inspiration to make The Gravel Garden.' Beth and Christo had gone to Dublin to give talks and took the opportunity to visit friends and gardens, including travelling to County Clare.

Beth used hosepipes to plan the shapes of borders during the creation of the Gravel Garden.

Here they were taken to The Burren, a rare example of a wild landscape formed of 'karst' hills – large limestone boulders with Mediterranean and alpine plants growing in the cracks. Beth wrote in her notebook: 'Everywhere the limestone lies in bands, fractured and weathered into curious wavy cracks and fissures. I hung over a little bridge looking at one of the few small rivers showing above the ground – most of the rainfall disappears into underground rivers. Suddenly it gave me an idea for the design of my Gravel Garden, so I made a little sketch to remind me of the gentle outlines with promontories here and there concealing little bays, and tiny islands of sedge with outlines of *Iris pseudacorus* – and odd large boulders – all this could be made with small groups or isolated plants like lavender or ballota for the boulders – grasses instead of sedge. It interested me very much.' Later she added a note to the page – 'My memory is failing! I thought it was New Zealand that I looked down on to a dried-up river bed!'

Beth and David Ward survey the planting laid out for the Gravel Garden.

The real work on the new garden started later in 1991. 'Today after lunch I made a start on the new Gravel Garden,' she wrote in her diary in the evening of 13 November. 'Keith [Page] had laid out all our yellow hosepipes to straighten them. I collected about eight border forks to fix to the ends and an armful of split canes . . . I began with the walk from the new car park leading to the nursery – to the left of the garden, straight ahead. After months of fretting about it inside I found that working it out on the ground I was enjoying myself. (Just like Chelsea – over anxious waiting for it to begin.)'

The idea of using hosepipes to lay out her beds was not new to Beth. In her *Notebook*, she had described how she regularly used them when enlarging beds in various parts of the garden.

Beth's plan was to keep a main path winding through the centre with tributaries leading off – one leading to the house, another to a cistus growing in front of the

The Gravel Garden in the early days.

leylandii hedge. She considered adding another further along the border. They were to have seats facing west more or less, with the backing of the hedge, where visitors could enjoy viewing the garden. David Ward took photographs of the area as Beth experimented with the hoses on the layout of the new garden. She called it 'Chatto designing in situ' and noted that she found the light and cloud formations helpful as well. The work had begun about two weeks before when Gerard Page sprayed the grass with Roundup® weedkiller. It soon turned a yellowish umber colour which Beth found quite attractive with the surrounding autumn colour. 'It is more helpful to me', she commented, 'than if it were a rich emerald green as it is in the car park.'

All the time, she had in mind a dry river bed like the one she had seen that summer in Ireland, with inlets and islands that were to be planted. The whole area was eventually 'mulched' with gravel, as were the walkways which represented Beth's dry river bed and its tributaries. She considered using different-sized gravels and planned to go to the gravel pit and see if the various heaps inspired her. In just over a month, the layout was complete and in the unseasonably mild and drizzly December weather, the first sod turned on the new Gravel Garden. The soil was so dry and compacted that Gerard needed a subsoiler to break through, taking great care to avoid all the service cables. 'A day to mark!' she noted that night.

Beth's Gravel Garden was to become – and to many still is – the most famous dry garden in the world, with its notice proudly proclaiming that it has never been watered other than by rain – in the driest county in Britain. This so nearly was not the case. After two particularly hot summers, Beth and David Ward were sitting on a bench in the garden one evening with Beth despairing of the wilted condition of many plants. She felt that she had to give them some water if they were to survive. Ward, usually the most cautious of advisors, pleaded with her not to. She didn't. It would be convenient to say that it rained the next day and saved all the plants. It didn't quite happen like that but rain did come a few days later and the garden's claim to fame stands firm.

The popularity of the Gravel Garden prompted the idea for another book to tell the story of its creation. When the contract arrived from the publisher Frances Lincoln, it asked for her to deliver 70,000 words. 'It seems to us it will be an enormously fat book with Steve's pictures which will be such an important part. Can one write 70,000 riveting words on one simple subject? Too many plant descriptions – even "poetic" passages can become soporific,' she wrote to her editor, Erica Hunningher, in October 1998.

In the end, the main text was shorter. As always, there were disagreements over both the editing and the design. Beth and Erica favoured a traditional treatment rather than, as Beth thought, 'unnecessary bands of colour down the sides of pages, and between photographs, plus turning the section headings on their sides so that you would have a crick in your neck to read them.' Others thought differently. But what did not change was Beth's devotion to and praise for Erica, by then a freelance editor, who, Beth wrote to her agent, 'rarely alter[ed] my words . . . rearranging sometimes a sentence, sometimes a paragraph, like threading beads of different sizes to make a graded necklace.' Praise indeed from someone who wrote so beautifully herself.

TRAVEL NOTES 5

Round the world with Christopher Lloyd (1989)

IN SEPTEMBER 1989, Beth left the UK for a month-long round-the-world trip with Christopher Lloyd, visiting New Zealand, Australia, Canada and, finally, Minnesota. She had been invited to speak at a symposium in Melbourne but had initially turned it down, not wanting to leave Andrew for so long. Then came a phone call from Christo to get her to change her mind. It would be such fun to go together, he said mischievously, he would be her companion. Despite having turned it down, Beth reluctantly agreed to see if the organisers were still interested in her coming – which, of course, they were.

For the first time ever, she was leaving Andrew for a month. 'It hurts to say goodbye. He always looks so vulnerable,' she wrote in the first of the four new red Silvine notebooks bought especially to record the trip. But the fun began as she arrived to meet Christo at the airport. First class flights had been arranged and Beth made her way to the lounge, where she found herself surrounded by 'exotic Arab ladies, expensively dressed [with] flowing robes, glittery jewels and flashing eyes'. Shortly after, Christo arrived, 'bustling in', Beth remembered, 'with his little holdall which he [took] around everywhere with him . . . at least 50 years old.' Out of this, Christo produced a brown paper bag with sandwiches for the two of them. 'He couldn't bear to leave behind his homemade bread and delicious home-cooked ham . . . so, in case I'd be hungry, he brought this to refresh me.'

They flew first to New Zealand, where she and Christo were each to give two lectures. But the real excitement was when they were taken out into the surrounding countryside and spent time botanising, seeing acaenas, hebes and many grasses and forms of carex which Beth grew in Elmstead Market but could now see growing in their natural conditions. Their host, Gordon Collier,* a well-known landscaper, organised their time. Visits to some of the best local gardens had been arranged during their brief time on the North Island as well. At a visit to Beverly Mcdonnell's garden in Auckland, Beth thought it curious to see forget-me-nots on one side of a path with strelitzias on the opposite side. On the South Island, she marvelled at the west coast road and shoreline where rocky banks fell almost to the road's edge 'garlanded with flowers . . . blue echium, mesembryanthemum, yellow daisy flowers.

Next came two weeks in Australia for a garden design conference attended by 1,400 people from across the world. Other speakers included John Brookes, Martha Schwartz,* Peter Thoday,* Anthony du Gard Pasley,* James van Sweden, Tom Wright* and Penelope Hobhouse.

Once again, Beth's notebooks are a mixture of plants and places seen, people met, clothes worn and meals taken.

MELBOURNE

After a picnic lunch we returned to hear Martha Schwartz – she was fantastic. Young, with spiky black hair she looked as if she had cut with nail scissors! She creates the most amazing urban architecture and designs. Has absorbed classical principles then thrown them away and proceeded to make them completely uninhibited and incredibly imaginative. She showed us murals in a prison entrance, a swimming pool, a shopping complex. I shall never look at railway lines with crappy sidings again without seeing her designs.

Marion Blackwell* followed – another brilliant woman, SW Australian, she has a fascinating accent, sharp and clear and fast. Both brilliant scientific knowledge and with a love that amounts to reverence for awesomely and poetically photographed natural features – deserts, mountains, river beds, rocks, soil, plants in all kinds of situations, showing all kinds of adaptations. It was a privilege to sit and listen to her. Both these two women have pushed out the walls of this conference hall and let in light and fresh air.

Sunday 8 October 1989

Our big day – up early to shower, shampoo, etc., wondered why my breakfast was late on this special day. Lucile was in the foyer to drive us to the hall. Immediately we were besieged by people, so friendly and enthusiastic and encouraging, saying they were waiting for us to speak. The first speaker was Peter Valder,* a very entertaining Australian. He talked about the woodland and rhodo garden first begun by his grandparents.

Next came dear Tom Wright of Wye College. He is such a nice man, modest, friendly and natural with a good sense of humour he looked good on the platform, dependable and solid with his broad shoulders and handsome leonine head. He spoke well, wittily and with lots of fresh aspects on what could have a mundane subject – garden maintenance – with very good illustrations.

After coffee Penny Hobhouse mounted the stage, her voice mercifully restored. I would have felt at a strong disadvantage in her place. However her script was immaculate and she did not deviate from it. Her delivery was clear, slowly and carefully enunciated. She informed us that it was to be a search for beauty – and that meant what was beautiful for her. Australia did not really come into it, not being of the right time in history. There was not a moment of self-doubt, not a flicker of amusement (did not raise a single titter). Her slides were without exception beautiful but every one represented great families of the past, of great wealth. We had great gardens of Italy, French castles including for the umpteenth time Versailles, finally descending to Lady Salisbury of Hatfield House, Gertrude Jekyll, and the interior designer David Hicks.

By now my mind was racing. It was my turn after lunch. Could I drag my thoughts away from these three speakers – all impressive – and concentrate on my opening remarks, still not fully realised.

I spent the first part of my lunch break with Graham, our chief cameraman and producer of the video that will eventually result from these two weeks together. We had to work out where I would stand so that the spotlight would catch me enough for filming. Martha Schwartz had been so put out by 'those damned overhead lights that blinded the speaker!' she threatened to shoot them out! Some were reduced. I found Peter Valder, Tony du Gard Pasley and John Brookes all very supportive. John put a hand on my shoulder and whispered, 'We've just had the Queen, Beth, you go on and be yourself. That is what they are waiting for.' He said it with such a naughty look I burst into laughter and gave him a hug. Just the break of tension I needed. John has emerged as a good leader during these past days.

Back to the conference. I went to the loo, tidied myself up then sat and looked at my possible list of openings. Then put it back in my bag and didn't look at it again.

Back in the hall people were saying they had come especially to hear Christo and me. Tony told us this last day had the highest number of bookings – over 1,400. So I went on to the floodlit stage feeling happy and confident, my thoughts flowed in sequence. Among my introductory remarks, I thanked the Australians and Tony in particular for giving me perhaps the happiest two weeks of my life. Then a pause, feeling I might be being unfair. I said I hope my husband does not hear that, which brought a gale of laughter. The rest was easy – just had to bear in mind I did not overrun since Christo was next.

At the end I had huge applause and obvious joy from my team of fellow speakers most of whom stood clapping as I came down the stairs. It was a great moment but my knees were knocking, this time with relief. But I was not free – many people bought books to sign.

So finally it was Christo's turn to be the last speaker of this very exciting conference. He too was a great change from the other speakers, using only a few headings for his introduction then leaving the podium and standing on the edge of the platform. He held the audience captive as he nattered away as easily as if talking to two or three, a bit naughty now and then but everyone loving it. He ended his talk called 'Softening and strengthening the garden with plants' with a beautiful slide of the huge fruit, seed heads, berries and autumn leaves that I made for him in the Great Hall last October. It remained on the screen whilst George Seddon* made the summing up speech.

What a nerve-wracking and exciting day. I was surrounded by people with books to be signed or just to tell me how much they enjoyed my talk. Finally we met again, all the speakers except Penelope, at an Indian restaurant for our final meal together. It was a crowded noisy room with everyone's pent-up emotions, anxieties and fears all resolved in this happy relief that it was over and had gone well.

After all that excitement, Beth and Christo stayed on for a couple of days' rest and recreation which included going to see a performance by the French mime artist Marcel Marceau. They both found the silent two-and-a-half-hour performance 'spell-binding'.

They then left for Vancouver, where they stayed with friends of Christo's, and on to Toronto for another three-day conference entitled 'Great Private Gardens'. While they were there, they were rushed around Toronto by various friends for an exhausting programme of entertainment and garden visiting. 'We both collapsed on to a big mattress-like bench, and Christo admitted he was tired too and his feet hurt. We are both country bumpkins not able to take hard city life,' Beth wrote that evening.

There was one free night where the pair had a quiet dinner together in the hotel. Beth wore a white silk shirt set off with her Chatto cameo brooch. The meal was delicious, Beth recalled, and the wine slipped down easily. The pair of old friends talked easily, discussing the day and comparing Toronto with Australia. Christo said he knew of no one else who would not have irritated him if they were together for three weeks. Beth laughed and reminded him they still had a week to go. 'It was a lovely evening,' she ended that day's entry.

The next day Beth and Christo's transport to the conference failed to materialise and they were left 'sitting like children, waiting for our minder'. A taxi was called and the pair arrived in time to hear Pamela Harper,* whom Beth had met on an earlier trip to Portland, Oregon, read her paper on the northern California garden of Harland Hand. Beth was not impressed.

Pamela had very good photographs and her talk and sequences were well put together but she has a flat-sounding hard voice. It's an embarrassment to me but I cannot like that woman having met her now several times. Certainly she shows not the least sign of liking me.

The next speaker was Frank Cabot, who had properties in the St Lawrence River area and in New York. He talked about the family home in Canada, and how he had developed the garden over the past twenty years.

Not one to do things by halves, it was mind-boggling as three carousels at least (80 slides each) were hurried through in rapid succession as Frank read his lively and amusing account, even giving an erotic flavour by drawing attention to a certain architectural erection – and then giving his psychiatrist's interpretation of his love of large leaves. (Did he develop that by any chance after his visits to my gardens many years ago?) It was a dazzling performance by a larger than life character and I thoroughly enjoyed it – even though I should imagine not one soul in the hall could emulate it or learn from it, since it almost reads like a cookery book (only reinterpreted) – 'Take 100 gardeners . . .'

After Toronto, the pair were off to Minneapolis, Minnesota, where they were met by C. Colston 'Cole' Burrell,* another friend of Christo's and then Curator of the Minnesota Arboretum where Beth was to speak.

That evening we were invited out to have a Thai meal with Cole and a friend – another very nice man. A bit older than Cole I would guess, very sensitive face. He is a social worker in the big hospital in St Paul, the attractive old town across the river that is twinned with Minneapolis. We asked him about his work and he told us much of it connected to the suffering of people either with Aids, or those who have lost loved ones with Aids. A harrowing job. Must have a strong and compassionate character behind that sensitive face. However, we did not have a gloomy meal – conversation flowed.

At the end we ate little sugary sweet cakes [fortune cookies]. Inside them were mottoes. Mine read, 'You have a quiet and unobtrusive nature.' I found this so far removed from the truth that I could not bring myself to read it, but was speechless with laughter at its inappropriateness, I handed it to Christo, who collapsed with laughter. When asked what his motto said, it appeared he had eaten it! More helpless laughter. So silly – but I hadn't laughed so much it seemed for years.

We then went to a concert of baroque music which ended with Mozart's Prague Symphony, the first and last movements so joyous they seemed to encapsulate and put a wonderful finishing touch to the whole month that has been so happy and exciting. We all went back to the hotel and had a last round of whisky – one of Christo's special malts – together. What a pleasurable experience it has been to meet these two friends in almost the middle of the North American continent.

14
Family and friendships

New year's eve always made Beth reflect on her life. It was now usually spent alone, and on 1 January 1991 she wrote about the winter night. 'Last night (of the old year) must have been full moon – up in the night to trip to the bathroom, I loved to see the bright light falling through the west facing bedroom windows, lovely patterns of shadows and shadowy furniture. Stood at my window, the garden almost as visible as day – and yet without colour – I never go out to see it in the moonlight – why not – partly because the moonlight is never so bright it seems as when one starts shivering in one's nightie at some unearthly hour.'

This particular year, 1991, started well. Christo rang to say he had booked 8 July for Glyndebourne and reminded her that he would be seventy on 2 March. Would she come and help make eats for his party? Of course. There were other pleasures as well. She brought in one of her large pots of jasmine and stood it in its place by the dining room window. Each year she looked forward to bringing it in and watching the buds open slowly over the coming weeks. She also picked the first blossoms of *Hamamellis mollis*. The final task was to take her big laundry basket to collect all the Christmas cards. She found it hard to throw them away, since, to her, they represented so many friends from near and far she had heard from over the Christmas period. Although she always insisted on opening her cards and presents in front of her family, she was happy to clear them away on her own. She would open each one and read it again before consigning it to the dustbin, feeling a little guilty not to find a way of recycling them to a charity. Just a few she kept back because she felt they were too good to throw away. That year one was David Ward's, a painting of an iris, and another from Romke van de Kaa, always made by him with leaves and bits and pieces from his nursery.

But Beth's peace did not last for long. One Sunday, she was horrified to be confronted by a fishing club, men and rods all around the reservoir that formed a border between the gardens and Hans's land. She felt, she later wrote, 'sick to her boots'. 'Hans knows surely how much I need privacy on Sundays. The public are watching my every movement all the week. Sundays are doubly precious because I can be truly alone, unobserved in the garden.' That afternoon, Hans made one of his regular visits when Beth was having tea with Andrew. Walking him towards the reservoir, she told him how upset she had been, having twelve men watching her. After Beth explained that she dreaded this kind of invasion of her only peaceful day of the week, he agreed to put a stop to it.

Another weekend, Beth's peace had been ruined by two lads on motorcycles racing round some adjacent land. She fought with herself not to feel mad about what felt like a blatant invasion of her and Andrew's privacy, destroying the peace and atmosphere of the little Wood Garden. 'These are the kind of things', she wrote later in her diary, 'that make me feel sometimes that I will have to find a little cottage somewhere else but where? Where could I be so happy, feel so at home as I do here? However beautiful it might be – and peaceful – how could I belong as I belong to this piece of land?'

This was possibly the only time she even briefly considered moving away from White Barn House. For most of the time, the weekend closure did give them an escape from the public gaze. From its inception, the nursery and garden had closed on a Sunday to give them some privacy. From November to March, it also closed on a Saturday. This did not deter some keen visitors. One winter Saturday morning, Beth looked out to see two women poking about in the nursery. Challenged, they feigned ignorance that the nursery was closed that day, although, noted Beth, they would have had to have passed through three gates telling them so. Although one was visiting from a long distance away, Beth did not relent but sent them away. (Eventually commercial pressures prevailed and by the start of the new century Beth was forced to open on Sundays.)

The weekends also were when she was able to see her family. Beth and Andrew now had six grandchildren. Diana lived in Nottingham with her second husband, David Peacock, and children Julia, Daniel, Lucy and Emily, while Mary was closer in Southwold, Suffolk, with Alastair and their sons, Thomas and Jeremy. Beth's

Beth with four of her grandchildren, (left to right) Daniel, Lucy, Julia and Emily.

rise to fame and the increased demands on her time meant that she had almost always had to put work before family. She visited as often as her commitments allowed and the two sides of the family came together when they could, but her relationship with her daughters was not always easy.

Within a few weeks of the fishing club incident, Hans – or 'Mr P' as a tight-lipped Beth was now calling him in her diary – caused more problems by announcing that he would not be allowing several of his winter pruning staff to move over to the nursery as they usually did to help out with packing. This was not the first time he had done this. Beth became enormously stressed and there was a face-to-face row in the office. She was so angry she threatened not to take any of his staff but to make her own arrangements in future. Eventually it was agreed that, while that year was a lost cause, a formal contract would be drawn up for Hans to sign, to avoid the same problem next year. Having known him for so many years, Beth had her doubts as to whether he would stick to it. 'We shall see,' she wrote that night. She was, she added, 'drained by all this hoo-ha'.

Another long-standing problem to do with rights of way with a neighbour – not Hans this time – prompted Beth to suggest they try to make legal the arrangements they had discussed and signed together a year previously. She drafted a letter for her secretary to type, adding, 'if he has further comments or objections, please get on with it!' but asking Rosie to put it rather more tactfully than that. When she showed the typed letter to Beth, Rosie quoted one of her favourite observations – that it was like trying to 'saddle a nightmare'. 'She says the same thing about Mr Pluygers,' wrote Beth that evening. 'She's right.'

THERE WAS NO LET-UP in the pressures on Beth. Every year, as the season got going, a stream of photographers and camera crews arrived. They were rarely refused since every visit meant more publicity and hopefully more visitors to the garden and nursery. The downside was that everyone felt a little stretched, especially Beth, who was always in demand. She was rather bewildered when a journalist from *The Sunday Times* magazine, came to interview her for their 'Day in the Life of . . .' series. 'There was a long discussion – some of it fairly silly – "What did I have for breakfast?"! "What hand cream did I use?"! I am wondering if I should not give so many interviews?' she pondered. 'So far I have been fortunate not to have my words or activities misinterpreted.'

Unfortunately, this is just what happened shortly afterwards. The aptly named 'Weasel' diarist in *The Independent* newspaper's magazine wrote a vicious piece mocking Beth's 'Day in the Life of . . .' story and deliberately distorting her comments. Having told readers that Andrew was fourteen years older than her, 'so his day goes at a slower pace', she went on to mention that she woke up at 5 a.m., and that he brought her breakfast in bed. 'I hope she doesn't demand breakfast immediately from the poor old slow-coach,' sniggered the Weasel, who carried on in this vein, reinterpreting with an acid tongue every comment Beth had made about her day. The piece ended with an appeal to the *The Sunday Times* magazine 'to invite Andrew Chatto to write his own "Day in the Life of". It should make compelling reading.'

It was Beth's brother Seley who leapt to her defence, writing 'a very pained letter' to Alexander Chancellor, then editor of *The Independent* magazine. In response, Chancellor wrote directly to Beth claiming that most readers would have seen that it was simply intended to show what a contrived and misleading form of journalism the original *Sunday Times* piece had been. He did concede that it was 'susceptible to being differently understood' and for this he was sorry. The apology was rounded off with the admission that he, his wife and his ninety-year-old mother were all great admirers of Beth, which, somewhat contritely, he admitted made him 'all the more regretful that pain should have been caused to either you or your husband'. Beth showed her gratitude for Seley's indignation by taking him to a piano recital in Framlingham Church in Suffolk that evening.

Inevitably, during May, her thoughts would wander to what was going on at the Royal Hospital site in London. 'Today is the end of Chelsea preparation,' she wrote on Monday 20 May 1991, 'all tidied now and waiting for the visits of the royals and VIPs. I did enjoy my 11 years but am very thankful not to be there now. We simply could not cope. Everyone as busy as bees here today. A lot of visitors – a nice day, quite warm but dull until the evening. It's a lovely feeling that things are improving here all round. I need the time to be here.' Later, Beth added a note at the top of the page saying that she had not written her diary again for three months because every day was too busy. Unusually for her, she felt too tired – 'or lazy' – by the end of the day to be bothered to write. The diary writing stopped for a full five months. She did not catch up until after another visit to Shrubland to recuperate from stress.

What prompted her to start writing again was yet more stress and something of a Black Friday for Beth. Going through the post in the office on 1 November 1991, she came across a letter of complaint from a customer about some small plants. The writer also enclosed a damning article in *Gardening Which*? magazine showing that the nursery was in the bottom end of a list of suppliers. Judged for packing and delivery standards, Blooms of Bressingham had been given five stars, Sandra Bond of Goldbrook Plants in Suffolk (Beth's 'bête noire') had four, while the Beth Chatto nursery only had two. Beth felt sick with shame, and immediately wrote a conciliatory letter to the customer enclosing a refund. She had been aware for weeks that there were fewer orders coming in but had put it down to the drought and the recession. 'I must say I feel sick to my bones that my work has been knocked to the bottom of the class. I went to tell the packing staff more in sorrow and despair rather than anger that we must put a stop to this.'

It transpired that Rosie, Beth's secretary, and David Ward had known about the article since the summer but had decided to keep it from Beth as they felt she was under enough strain. While Beth could see that it was considerate of them, she was upset at having been kept in ignorance for so long while the rest of the staff knew about it. The garden journalist Anna Pavord, in touch about another matter, tried to console her. 'You were obviously shaken by the *Which*? thing . . . I can't say I took it too seriously. The sample is so tiny . . . one learns to take these things with a pinch of salt, but no one could put more effort into the nursery than you do. *Courage, mon brave*.' To cope, Beth disappeared to the vegetable garden, her 'place of calm', where she planted some garlic and picked armfuls of flowers. The following Monday she went to the local signwriters and ordered the following plaque: 'Please do not pack any plant you would not be happy to unwrap.'

There were no more such dramas after this and business picked up without any lasting damage to the nursery's reputation. A couple of years later, Beth was delighted to be told that two orders had arrived from Mr Goldsworthy, head gardener at Buckingham Palace. One was for the Palace and the other for Clarence House, the London home of Prince Charles. They also received orders from the Royal Parks for Regent's Park, St James's and Hyde Park. Beth was thrilled to know that they came to about £5,000 in total – 'a sizeable whale', she declared, 'from the tiny shrimp we had a few years back'. Goldsworthy had written listing plants he wanted to renovate the original grey border, which had been made by Beth's old mentor, the 'Silver Queen', Pamela Underwood. Beth thought it could do with a bit more variety and wrote back with suggestions. To her delight, he accepted them and came down to collect the plants himself and have a look round the gardens.

BETH WAS NOW not only supplying plants to the royal family, she was also in demand at some of the grandest houses in the country. Her skill as a gardener, writer and presenter, but most of all plant expert, had brought her into a world unimaginable when she had been a police constable's daughter in rural Essex. In 1993, she accepted an invitation to stay at Hatfield House in Hertfordshire for the weekend as the guest of Mollie, Marchioness of Salisbury. Beth had been invited to speak at the Festival of Gardening in June that year. She accepted, waiving her usual speaker's fee of £500. 'I admire very much what you have done to enhance the gardens and will happily do what I can to support you in your great task of preserving the house also,' she replied to Lady Salisbury.

She was nobly entertained during her stay and impressed by having had her suitcase unpacked by the butler. 'Every table and surface held beautifully arranged flowers from the garden (by Lady Salisbury), and pot plants from the greenhouses. The effect was enchanting, because of the setting – large proportions, panelled walls, fine paintings. Lamps large and small, soft but clear coloured furnishings. Two well-behaved dogs completed the perfect country house atmosphere. One a tiny smooth coated little creature curled up like a cat on Lady S's lap. I was introduced to Lord S, a large handsome man but finding getting up and down rather difficult. A low table beside the sofa on which I sat was covered with small glass pots, each filled with different flowers. They reminded me a little of the lunch table specimen vases that were made at Baron Philippe's. One held flower heads of the shrub that smells of stewed prunes (so Christo and I think, but no one else here could recognise it as that, merely that it was a spicy scent). It filled the space where I sat.'

Her friendship with Christo had become a bedrock in her life, despite his teasing about Andrew, her 'brown garden' or, cruelly, over her despair at the sudden death of one of her earliest staff members, Madge Rowell. This felt to Beth as though she had lost her closest friend, someone who had shared all the best and worst of the garden and nursery. 'Her straightforward common sense, warm heart, and unique sense of humour', Beth later wrote, 'all often saved the situation, and restored me when I had lost faith in myself.' And just as Beth learned not to mention Graham Thomas in front of Christo, there was more than a hint of jealousy in his teasing of Beth about her constant references to plants as 'Cedric's this . . . or Cedric's that'.

More than once he overstepped the mark with remarks about Andrew, calling him 'St Andrew' and 'her ball and chain'. He often said Andrew was holding Beth

Beth with Madge Rowell, who died unexpectedly in 1992. 'I feel I have lost my closest friend, sharing all the best years of the garden and nursery.'

back from lecturing and travelling more, especially when she turned down trips with him such as one to America in 1993. He even suggested that she should leave Andrew. After one particularly acerbic comment, late one evening when she was staying at Great Dixter, Beth was on the point of packing her bags and going back to Elmstead. A few private tears and some deep breaths later, she decided not to go. The next morning, Christo made a rare apology, admitting that perhaps he had gone 'a little too far'. He was contrite as well about the trip to the US, writing later that he appreciated that, given Andrew's health problems, 'you would not be easy in your mind either in the run up or, perhaps, while actually away.'

To others, Christo constantly sang Beth's praises. In contact with an American publisher about her, he said, 'I know of no more honest writer. She describes as

Beth during one of her many visits to Great Dixter, enjoying the planting style that was so different from her own at Elmstead Market.

she sees, imaginatively but never over the top; always so as to make me murmur an echoing "this is as it is". Everything is from direct observation and experience. She is never woolly or off the point. Beth Chatto knows how people think and behave and pinpoints our foibles and weaknesses with common sense and understanding.'

Whereas Christo made few trips to Elmstead Market, with gaps of up to eight years between visits in the early decades of their friendship, Beth continued to go to Great Dixter regularly. Over many years, she was always with Christo for his birthday in March. Life at Dixter was so different from that at White Barn House, where caring for Andrew now took a great deal of Beth's time and living in the centre of the nursery and garden meant no escape from business anxieties for either of them. Great Dixter was another world. Christo's legendary hospitality meant that the weekends included an annual visit to nearby Glyndebourne, while back at the house there were always stimulating guests. (Though the rambling stairways and guest rooms at the 500-year-old house in East Sussex presented their own problems. While it had been sympathetically extended with original materials in the early twentieth century by Edwin Lutyens, modern amenities were still in short supply. Visitors in one of the four guest rooms at the top of the house risked bumping into each other on the way to and from the only bathroom nearby.)

Christo loved to surround himself with interesting people from all walks of life. One weekend, Beth arrived to be told that he had invited Beatle Paul McCartney and his wife, Linda, whose country home was a farmhouse at Peasmarsh, a village less than five miles away from Dixter. He insisted that Beth cook the meal, since the McCartneys were famously anti-meat and he 'didn't do vegetarian'. Nevertheless, he was happy to go shopping with Beth and her meal of stuffed aubergines, salad and herbs from the Dixter vegetable garden was a suitable success. Beth already knew George Harrison from Chelsea and now she made firm friends with Linda McCartney, staying in touch until Linda's early death in 1998. She kept their notes and their Greenpeace Christmas cards decorated inside with love and hearts.

Given the closeness of Christo's friendship with Beth, it is quite surprising that when, in the mid-1990s, the idea was first mooted for a book of letters with him, the publisher Frances Lincoln's suggestion had been that they should be between Christo and Rosemary Verey, then the upper-class queen of garden design in muted colours. It was Beth and Christo's agent, Giles Gordon, who immediately rejected that coupling, correctly seeing Beth and Christo as the perfect combination. After Gordon broached the idea to Beth in early January 1996, she wrote back enthusiastically, never knowing she had been 'second choice'.

'You crafty old thing! Such a clever trap because I find the idea very attractive. Alarming too, the fear that I might not have enough time, especially in the very busy months, but then there's also no shortage of copy . . . Dear me, with so much else going on here I feel I am about to leap over a cliff edge, but it is an exciting way to start the New Year. I hope it may turn out well for us all.'

Beth was soon jotting notes on scraps of paper on what she might write about, 'Next letter (1) Refer to wooden frame put up right. Hall, comparison with Tea House, (2) Refer to new grass – *His Health*!' Christo set the ball rolling with a criticism of Beth for not watering the Gravel Garden. 'Obviously,' he wrote, 'I'm not such a moral person as you. But when I see plants suffering, it's like having a pet animal that you've neglected to feed. But an animal will touch your heart with mewing or whining. The poor plant cannot do that.' When she read this, she threw the letter down on her desk in a state of high dudgeon, marched out on to the nursery with thoughts buzzing around her head. 'If this is going to be the style of exchange, it begins and ends here,' she decided. But by the time she had got to the Wood Garden, she had calmed down and was intrigued by it. 'Right,' she thought, 'I want to write this book – you've given me the kick-start I need.' Her letter in return to him came with a covering note, 'So much is happening already in the garden. I hope next time I shall write about that but I enjoyed having a lead from you, a bit like a squib going off under my skirt.'

Quite soon, their publisher, Frances Lincoln, having seen these first few letters, was alarmed, saying: 'there's not much love lost between them', or words to that effect. While Beth was only too aware that Christo needed no encouragement to tease her about her views on ecology, she wrote to him agreeing with their agent Giles Gordon that they, or at least she, had not got the tone quite right. She felt she needed guidance on which bits sounded stilted. 'I have also in mind a sweet letter I had from Fergus [Garrett, head gardener at Great Dixter] encouraging me to accept the offer to do this project, saying "the wealth of knowledge being exchanged on a personal level means an informative and readable book. Christo wouldn't want to do this with anybody else." Hence I have tried to include a few descriptive passages – it is hard not to think of the person who may read the book – as well as of you – but neither do I wish to sound as if I am lecturing the world . . . My problem is that we are not writing just to each other. There is so much already experienced and understood between us which would be taken for granted if these were private matters. Quite how to strike a balance I'm not certain. I look forward to being at Dixter Sunday 4 March, when maybe you will unravel a somewhat confused Beth.'

Matters got worse as, quite soon, Beth took exception to Christo's many digs at her devotion to organic gardening, especially organic vegetable growing. His sarcasm riled her, while her dogmatism infuriated him. The great friendship was tested to such a point that a meeting was called at Great Dixter when the pair threatened to call the whole project off. Fergus Garrett was called in to mediate,

Beth's friendship with Christopher Lloyd lasted to the end of his life.

as publisher Frances Lincoln struggled to get the pair to agree on the right tone for their letters. Fergus had only recently become Christo's head gardener, but he had known both Christo and Beth for several years. He had spent a short spell working for Beth after he completed his training at Wye College, but it was not the right place for him. Beth was not ready for any head gardener other than herself, whereas Christo explicitly encouraged Fergus to become a better gardener than him. Both Beth and Christo knew that Fergus had a wise head on young shoulders and that he was the perfect person to arbitrate between them.

Eventually, after a lively debate in the Parlour, Christo in his armchair, Beth on one side of the fireplace and Fergus on the other, with one or the other of the

Christopher Lloyd and Beth with Frank Ronan, Miles Johnson, Anne Wambach and Monika Schuster at Glyndebourne in 2005.

warring pair stepping outside to cool off with a walk round the garden, agreement was reached to carry on with the book. It did not take too long to smooth things over, Fergus recalled, 'because they had the greatest respect and love for each other'. Things seemed to be back on track, except now it was Frances Lincoln who complained, saying she felt that they ought to stick to 'gardening information' instead of 'too much else (such as honorary degrees and opera)'. This time, Beth and Christo closed ranks and insisted on writing about what they wanted to.

'I quite accept', Beth wrote in heated tones to her editor, Erica Hunningher, 'that if we're not writing the way you and Frances Lincoln would prefer we may not be published by you (or anyone else!) But frankly we all feel . . . that there is a plethora of garden information about – are you really excited to pick up yet another book or garishly illustrated mag on How to, or You should be . . .! Possibly there's a too limited audience for our sort of chattering letters, but I think we need to go on – to let them develop a style and substance. Frankly I don't mind too much if they're not published (even though they do take time and effort on my part, and I enjoy writing them when I can find a quiet uninterrupted day). Christo is much better than I at tossing them off late in the evening, but then he is a much more professional, more practised writer than I am.'

Eventually feathers were smoothed and the writing continued with the talk of visits to Glyndebourne for opera and illnesses remaining. It became one of Beth's best-loved books. Christo was pleased with the title Erica Hunningher came up with, 'Dear Friend and Gardener' – 'not sentimental but warm'. 'Between Friends', another suggestion, was, he thought, 'a bit truncated'.

When a new edition of *Dear Friend and Gardener* was produced in 2013, there were mixed views about the addition of illustrations, since Beth and Christo's words were so evocative of their lives and gardens. Beth thought the addition of photographs, however good they were, was a distraction from the words. She also wanted what was thought by some to be an exaggerated amount of credit given to Andrew's research in the new foreword by Fergus Garrett. Fergus tempered her words, writing, 'fuelled by the studies of her late husband, Andrew Chatto, on wild plants and plant communities, Beth's garden and nursery, coupled with her words on paper, have changed the way we think.'

When it came to reprints of her other books, Beth had strong opinions about new cover designs and often disapproved of any 'improvements'. 'While the overall pattern is attractive,' she wrote of a proposed paperback reprint, 'it is anonymous, it represents general gardening, whereas we are specifically concerned with suiting plants to the situation . . . rather than putting them anywhere, regardless of their requirements.'

15
Falling leaves

Now in her seventies, Beth showed no signs of slowing down. With the instant success of *The Gravel Garden*, she was in demand as much as ever, reaching a new public who had never seen her Chelsea exhibits. It brought her a worldwide fame with its links to the effects of climate change far ahead of the public's understanding of changing long-term weather patterns. Most gardeners just saw it as a solution for dry beds without realising that, as Beth knew only too well, with the coming of more variable, and generally hotter, weather due to climate change, in the future it would not just be arid Essex that would suffer but gardens throughout the country and indeed the world

When she wasn't gardening, she was writing. Beth was asked to be a regular contributor to two magazines about to be launched that spring. She liked the idea but knew that time constraints meant putting it into practice would be difficult. 'I need to know more of what will be required and when,' she explained. 'I do not like writing from imagination or memory. It's so much better when I can be inspired by looking, feeling, smelling and touching – something always seems to produce a spark to write, something I couldn't imagine from memory.'

Beth was never reluctant to check and edit another person's writing about her or the garden, even when it came from the revered writer, and her best friend, Christo. In the mid-1990s, he asked her to look over a piece he had written about her for a book that later came out as *Other People's Gardens* (1997). 'I never wrote in the details that I had at first intended on the gravel garden, because by the time I reached it, I felt the reader would be getting sated with detail, but I don't really think that matters,' Christo wrote to Beth when sending her his copy to check. 'Selectivity is an art in itself. I believe I've said what I wanted to in the space I've allowed myself.' Five days later, Beth replied, later noting in her diary that she had

Even without Chelsea there was no let-up in Beth's life, with constant writing commissions and adjustments to be made to the garden.

made quite a few corrections; 'Since the account might last longer than the garden, I wanted to get the facts right! Most of it was very perceptive, heart-warming for me to read, but here and there he wasn't quite accurate. I made notes and underlined passages or lines in the typescript – with page numbers and note numbers all written in so I hope he won't find it too tedious to link together.'

There was also a constant stream of letters to be answered. She always wrote her replies by hand initially; some were then typed, by Rosie or later by the equally devoted Tricia Brett, but more often than not they were sent as personal notes. A sunny January day in 1994 saw her writing both a letter and a birthday card to an old friend. 'Poor [X],' she wrote in her diary that evening. 'She is very difficult

now. Difficult for herself as well as everyone else. How lucky I am still to be able to work at all the different things I enjoy, whether physical or mental. I am aware how much I need to be able to entertain myself in this way to preserve my health and happiness. I too sink into depression if I have no sense of achievement. Not always big new things are necessary – sometimes clearing up or sorting out is a kind of relaxation. But to be occupied and feel that it is good – that is the base of my equilibrium. Is that bad? Should I spend more time preparing myself for a less active life – absorbing more, listening more? Being lonely and helpless like poor [X] is unbearable to contemplate.'

Keeping occupied was not a problem for Beth in the 1990s. In addition to the writing work, there were many media requests. In 1994, she received two telephones calls from a desperate producer for BBC Radio 4's *Gardeners' Question Time*. A dispute over contracts had led the old team to leave the BBC en bloc, defecting to the commercial radio station Classic FM. This left the question master, Eric Robson, with no one on his panel. Would Beth go to Manchester to record three programmes? Then someone else from the BBC tried to talk her into several other things, including something at Hampton Court. She turned them all down. 'I do not wish to be a media personality and simply haven't the time.' Would she be approachable if they had a programme in the area? 'Not wishing to be utterly churlish, I said, yes, that was possible.'

Less than a month later, the BBC came back to her with another offer. Would she take part in a *Gardeners' Question Time* to be recorded in two sessions at Clacton, just ten miles away? This time she agreed. She already knew one of the panellists, Geoff Smith, but not the other, Bob Flowerdew, although his reputation intrigued her. 'We've exchanged letters. What a name! Is it real or has he just invented it to go with his charisma? He's a dedicated organic gardener. I think it may be worth going just to meet that young man.'

To help supplement the permanent staff in the nursery and garden, Beth continued to take on a succession of students, almost all foreign. Many, such as

Beth always loved growing vegetables, and this private part of the garden became a sanctuary from her hectic life.

For many years, the caravan provided accommodation for visiting students, not a problem in summer but tougher going in the winter months.

Yuko Tanabe in Japan, Doug Hoerr in the US and Peter Janke in Germany, went on to develop strong horticultural careers around the world but stayed in touch with Beth as lifelong friends. On arrival, as happened to Doug Hoerr, students would usually be billeted in Beth's ancient caravan tucked out of public view by her vegetable garden. Winter nights could be testing but the welcome was always warm.

Typical was that received by Hilde Sartori, a new young German student who arrived at the gardens in March 1991. Beth prepared the caravan, leaving home-made sweetcorn, leek and lentil soup on the stove and a warm electric blanket on the bed. Sartori returned to spend three summers with Beth and would often find small cards left to greet her alongside the ever-present home-made soup. Beth was

Beth with two visiting students, Bernard Trainor from Australia and Hilde Sartori from Germany.

delighted by Hilde, who kept the caravan 'spic and span – like a small child playing at house' – but also grasped Beth's ideals for good compost-making – 'the necessity to feel all the micro-organisms that teem in the soil, millions in a teaspoonful'.

Hilde might have been surprised to find herself working alongside a fifteen-year-old schoolgirl, Emily Paston, who had been working at the gardens for a couple of years as a Saturday girl. In 1991, Emily had brought Beth a school essay in which she had described a quiet moment in the garden when she just had plants and wildlife for company. Beth kept the essay all her life, adding a note that Emily always came in an hour before opening time and walked around the garden. 'She probably knows it, its changing moods and scenes, better than anyone else who works here,' Beth wrote at the bottom of the piece. She also added that she had offered her a full-time job as soon as she left school. 'We will be lucky to have her.' Emily did join and nearly thirty years later is one of the key staff members in the nursery.

Working for Beth was never an easy ride. Hilde was wise enough to ask someone if she wasn't 100 per cent sure what Beth wanted her to do. Beth set high standards for herself and others. There was, as always, 'the Beth way of doing things, and the wrong way'. 'She could', said one long-serving staff member, 'reduce grown men to tears.' A former volunteer also told of Beth leaving her a quivering wreck after a particularly acid dressing-down for something she had done wrong. Friends weren't immune from her comments either. Germaine Greer remembers being at the receiving end of a withering put-down on a visit to Miriam Rothschild, famed for her wild-flower garden. When Germaine wrongly identified some plant, Beth stingingly snapped, 'You surely know better than that?'

But this directness was based on an unparalleled knowledge of plants. Dan Pearson, now a celebrated garden designer himself, was only thirteen when he first saw Beth's stand at Chelsea. Conquering his shyness, he began a correspondence with Beth and went on to get to know her well. While he has said he always found her thoughtful and wise, she could also be direct and to the point. He remembered waxing lyrical about the joys of *Smyrnium perfoliatum* before Beth politely interjected that it was quite a seeder. He replied that it hadn't been with him. '"Just you wait," was her fast reply. 'Of course, she was right.'

Students also soon learned not to question Beth's ways. Staff and students fell into one of two groups – those who couldn't take the heat in the packing shed, and those who bit their lips, came up to Beth's exacting standards, and stayed with her for years, often until retirement. Where bonds formed, they were close and strong. Despite her toughness, Beth inspired a protective loyalty and devotion from her staff. This book could be filled with the paeans of praise from many who came in contact with her. Yuko Tanabe called her 'my English mother' and treasured finding notes from Beth in the student caravan with raspberries and calendula flowers for her breakfast. Peter Janke, who, like Beth, came to gardening from a floristry background, went for a weekend's trial and ended up staying for several months. Doug Hoerr's story has already been told. Beth's files are filled with letters

Beth receiving a Lifetime Achievement Award from the Garden Writers' Guild (now Garden Media Guild) from Alan Titchmarsh in 1998.

of gratitude and appreciation from former students who found working with her life-changing.

There were others with whom Beth's small acts of kindness stayed forever. Having met her first in the 1970s when he interviewed her as a young journalist for *Amateur Gardening*, Alan Titchmarsh returned many times. There was always home-made soup ready whenever he visited and special plants put aside for him to take home. 'She was always so approachable – nothing was too much trouble,' he said. In 1991, while studying at Hadlow College in Kent, garden designer Matthew Wilson and some fellow students visited the gardens and were disappointed to hear they would not be meeting Beth since she was ill. Finding a scrap of paper, they hastily wrote a 'get well' note and pushed it through her letter box. To their delight, a few days later they received a long handwritten reply from Beth saying she hoped they would return and meet her another time – which they did.

TV gardener and presenter Christine Walkden particularly remembered an event from May 2009. She had just finished filming coverage of the Chelsea Flower Show on television, while coping with a cancer scare and, worse, fear that the death of her father might be imminent. Feeling totally drained, Christine recalls waiting for a taxi home outside the showground when 'a small lady in a grey coat came up to me and said, "You won't know who I am but I really love the work you do."' Christine replied that she did know who she was and was thrilled to be told that the famous Beth Chatto enjoyed her work. 'I travelled home with the biggest smile on my face . . . Beth would have had no idea about my circumstances, but she made me feel like a million dollars that day.'

Although it may not always have felt like it to her staff, Beth made it clear throughout her diaries that they were always a top priority, with endless concern for their futures. Only one, however, called on Beth literally to get him out of jail. Beth was slightly puzzled to get a call one day from Michael Dale, an ex-student now back in Australia. Michael had worked for Beth in 1991 and Beth had taken to him immediately, sensing he was going to be 'a treasure'. They often worked together, Beth wrote, 'as though he has been with me for years'. When he called her in 1991, after the usual pleasantries, he told Beth that he was in trouble and had been given seven days in jail. Could she fax a reference that afternoon? He had been caught smuggling in plants and snowdrop bulbs. 'Poor Michael,' said Beth. 'He was unwise but not evil. I wrote something to the effect that I thought the fright he had had would keep him within the law, that it was not with evil intent, and a pity to put the stain of prison on his reference.' Four days later, she received another call from Michael in Melbourne. He had been released by the court that day with his prison sentence quashed; he claimed it was due entirely to Beth's letter.

LEFT Beth with legendary Dutch designer and plantswoman Mien Ruys, in Holland.
BELOW During a visit to Piet Oudolf (second right), Beth once again met nurseryman Ernst Pagels (with beret).

After the extensive world tour in 1989, Beth no longer wanted to do long-distance travel. She did make short visits by herself to Holland, where she saw Piet Oudolf at Hummelo and met the great Dutch designer and plantswoman Mien Ruys. Another meeting with Ernst Pagels was also arranged. Pagels was so taken with Beth that he later named one of his miscanthus grasses after her – 'small and elegant just like her'. She also managed a visit to Ewald Hügin's nursery in Freiburg, close to the French border. Hügin had worked for Helen von Stein-Zeppelin at Laufen. While in his nursery, Beth spotted *Sedum* 'Matrona', which Hügin had found in his garden. With her eye for good plants, Beth was one of the first to bring it back to Britain.

THROUGHOUT THIS TIME, Beth's biggest worry remained Andrew. Ill health dogged the latter years of his life. Visitors to the gardens rarely saw Andrew, but those who did found him shy and scholarly, always the polite gentleman. The staff thought highly of him and the guests who occasionally saw him remember his gentle courtesy while he asked them whether they would prefer an Amontillado or an Oloroso sherry. To his family, he was a loving and often witty father and grandfather who simply preferred his own company to Beth's hectic lifestyle. Although fascinated by plants, he only visited the Chelsea Flower Show once and left 'after five minutes'. He taught himself Russian to further understand the plant habitats of the Urals, yet never made the journey to see any of them for himself, preferring to do his research from books and maps rather than 'on the ground'. He taught his daughters about nature, politics and travel, always encouraging them to be non-judgemental. When they were young, he got them to make dens in the summer and roll snowballs in the winter. And he had a somewhat surprising fondness for the Rolling Stones.

Andrew was unable to speak for the last six months of his life and had to write everything down. He also needed constant oxygen. Even in this condition, those around him remember his perfect manners, kissing the back of the hand of any woman who visited him. On the day he died, 20 September 1999, he and Beth had shared a little joke when he slipped away in her arms. She wept quietly as she laid him back, marvelling at how peaceful he looked. Beth was, by her own admission, not easily given to crying. It came, she felt, from having had to shoulder years of responsibility and not wanting to burden others.

Before his death, Beth and Andrew had discussed whether or not they wanted to be cremated. Beth still had sad memories of Andrew's parents' cremation service decades earlier but felt that cremation would allow Andrew and later her to have their ashes scattered in the gardens. Soon after Andrew's death, the local vicar came to discuss the funeral with Beth. He knew that, although she had strong spiritual beliefs, she was not a regular churchgoer, and that Andrew considered himself an agnostic. While Beth had been brought up in the Church of England, she told him that she felt she could find God in the garden and in nature. However, when it was pointed out to her that some sort of a memorial to Andrew would be wanted at the church, Beth relented and agreed to a burial. It was, after all, where they had been married, and their daughter Mary too. He was buried at the peaceful little Elmstead Market church. The church was full at the service, including his many friends from the local pub, and afterwards hundreds of rose petals were scattered on his coffin as it was lowered into the ground.

After his death, it was found that the twelve binders of Andrew's tightly written research contained around half a million words. It was a challenge to know what to do with all his hard work, but a conversation with plantsman and garden writer Noel Kingsbury brought an answer. Kingsbury placed a notice in the Hardy Plant Society journal asking for help with the transcription of Andrew's notes. He received fifty-five replies and soon a shortlist of six experienced typists was assembled to do the work. The results were put online on the Beth Chatto website, together with Andrew's beautifully hand-drawn and hand-coloured maps of the areas across the world that he had studied. This gave Beth enormous satisfaction and in a small way made up for the disappointment that Andrew was never able to publish his research during his lifetime.

By now Beth's five grandchildren were growing up fast. Her grandson Thomas had died in 1995 aged fifteen, from complications resulting from his physical disabilities, leaving the whole family bereft. Always a proud – although rarely a hands-on – grandmother, she amassed a folder of school reports, birthday and thank you cards from them over the years. Occasionally, Julia and Daniel, her eldest grandchildren, would come and stay but because of the limited space at White Barn House it was rare for the whole family to get together except for special occasions. In the post-Chelsea years, until his death, Andrew's birthday was always celebrated with a Sunday lunch in the garden.

Diana continued to live in Nottingham with her family, while Mary was still in Southwold, Suffolk, where her husband, Alastair Marshall, was head of wine buying for Adnams, the local brewing, pubs and hotels group. Of the two girls, it was Mary who was most interested in plants and gardening, and she did the flower arranging for Adnams' top hotels, the Swan and the Crown. But while Beth admired what Mary did in her own garden, she was always reluctant to involve her in the gardens at White Barn House in a practical way. Mary later trained to be a psychotherapist.

In 2006, the family was devastated when Diana, at sixty years old, was diagnosed with early-onset Alzheimer's. Beth visited her when she could, at her home in Nottingham where she was being cared for by her third husband, Tony Shaw, on one occasion staying for over six weeks. However, so quick was the progress of the disease that by 2010 she needed to be admitted to a specialist care home, a painful decision for all the family.

While Beth had a wide circle of horticultural friends, for most of her life she had few close female friends to whom she could confide her anxieties or her joys. One of her oldest friends was Janet Allen, a potter she had known since the Benton End days, who would regularly drop over from her home in Suffolk and knew the family well. Although she was one of Beth's rare confidantes, she also knew Andrew and the girls well enough to be amused when she saw them becoming a gang of three and rolling their eyes when Beth launched forth on some topic, with whispered comments, 'Mother's getting going again . . .'

BETH HAD ALWAYS HAD male admirers, both before and after Andrew's death. Over many years, she received letters addressing her as 'Betty', her pre-marriage name, often from boys she had sat next to on the school bus from Elmstead Market into Colchester in the 1930s, reminiscing about 'the old days'. One came from Raymond

Pryke from Capistano Beach in California. He had seen Beth on a CBS programme on Eccentric English Gardeners. 'He was a very handsome boy,' she remembered, 'but I was very shy of him.'

The most persistent of her admirers was the American hosta collector Paul Aden, with whom she had stayed on her trip to the US in 1986. She had never forgotten the moment he had lunged at her suggesting they become lovers. She dreaded his later visits to Britain. Aden was a tenacious soul who had a posthumous reputation for 'acquiring' new hosta varieties and then claiming them as his own. For nearly twenty years, he was keen to 'acquire' Beth as well, as his partner and soulmate. This fervour culminated in him sending Beth a formal proposal in the form of a legal contract laying out what both parties would bring to the table – or, in this case, Aden's home at the 'Garden of Aden' in Baldwin, New York. All Beth's living expenses would be paid by Aden (except for her guests) and after his death she would be supported in the home. The agreement 're: Working & Living Practice starting when suitable' arrived already signed by Aden, with the general note that 'since the benefits and the character of the parties are exemplary, we sign below our intent and willingness to live up to and even go beyond the agreement for benefit of the other party.' With a rare note of realism, Aden did add a proviso: 'If it happens that the meeting of the obligation and joy to each other becomes strained and intolerable, then after a warning notice and general discussion, this arrangement and agreement shall be ended.'

Once Beth's anger at such presumption on Aden's part subsided, she gathered her thoughts on how to reply to such an offer. The rejection was a short typed letter, very much to the point: 'While I appreciate your offer of a home in the US I am far too deeply rooted here (now in my 80th year) to think of doing anything else.' The occasional letter or email arrived after that but was ignored. While she claimed she had great respect for his horticultural knowledge, what she diplomatically called 'his social skills' exasperated her.

16
The final years

The new millennium brought fresh challenges for Beth, especially commercial ones. The new car park groaned with cars and coaches as people flocked to see the old car park, now the Gravel Garden. And all these visitors wanted to be watered and fed as well. Staff banded together to suggest to Beth that the existing shop and catering arrangements had to be replaced. At first, she was against the whole idea – 'I'm a gardener, not a waitress.' Her old worries about the gardens becoming too popular and just a tourist attraction serving pots of tea with cake were resurrected. Eventually, with reluctance about both the expense and the addition of a large building within the nursery, just by the house, Beth agreed. The result, which Beth called her 'Tea House', successfully provided space for visitors to sit in while they ate, and also a separate room that could be used for events, talks and lectures.

The design grew on Beth. 'It forms a comfortable background seen from a distance, whether the Gravel Garden or the nursery. I think I have to be careful not to plant too much "clutter", so am thinking of large tubs which can be moved around or re-planted. Hopefully it will evolve!' These displays became the responsibility of Emily Allard, formerly Paston, the young Saturday girl, who was by now the lead propagator of, among many other things, Beth's precious tender perennials and succulents.

In July 2000, Beth received an invitation to the opening of the new organic vegetable garden at Audley End, near Saffron Walden on the other side of Essex. The walled garden within the grounds of the Jacobean house, now owned by English Heritage, had been created and was to be run by the Henry Doubleday Research Association (HDRA, now more commonly known as Garden Organic). Doubleday, coincidentally also an Essex man, had been a nineteenth-century Quaker who pioneered the use of comfrey in the garden. The highlight of the day was the official opening of the garden by the Prince of Wales, the organisation's Patron since 1988.

Beth had been a member of the HDRA for many years and as the horticultural star of Essex she was a natural to add to the guest list. She had met most of the royal family at the Chelsea Flower Show but had never been introduced to Prince Charles – although it had done her reputation no harm when one year he was photographed walking around the Chelsea showground holding an Unusual Plants catalogue in his hand, with its distinctive leaf outline design for all to see.

At Audley End, somewhat surprisingly Beth was not in the formal line-up to be presented, but she was spotted at the end of the ceremonies by Jackie and Alan Gear, CEOs of Garden Organic, who called her over and introduced her to Prince Charles. They had, Beth later told me, an easy informal little chat. When Beth got back to the nursery, she found out that the next day David Howard, the head gardener from Highgrove, the Prince's country home, had driven across to Essex and bought several plants. Much to the delight of the staff at the nursery, Howard had told them that Prince Charles insisted on the visit after his meeting with her.

Beth was soon to meet Prince Charles again. Unbeknown to her, her staff had been lobbying since 1999 for her to be awarded a royal honour. Finally, in 2002, a letter arrived from Downing Street offering her an OBE. The normal procedure is for the invitee to sign an acceptance letter and then stay silent about the award until it is announced by the Palace. This was not quite how her staff wanted it to be. With no small degree of deception, Tricia Brett, Beth's secretary at that time, slipped the letter between some others, folding it over and saying that it was something to do with a birthday surprise so could she just sign it without looking? This went very much against Beth's normal way of operating where nothing would go past her without a thorough inspection, but this time she acquiesced. When the day of the announcement arrived, she was called into the tea room where all the staff were waiting for her. The surprise was complete when David Ward gave her the letter of acceptance that she had unwittingly signed a couple of months before.

Once the excitement had calmed down, there was the trip to the Palace to look forward to. Beth knew that the investiture would not be performed by the Queen, who was on a trip to Canada. She secretly hoped it would be the Prince of Wales, and it was. Once again, they chatted easily, and Prince Charles asked if Beth had

Beth was delighted to receive her OBE for services to horticulture from keen gardener Prince Charles in 2002.

been to Highgrove. When he heard that she had not, he promised to organise an invitation. Time went by and Beth's secretary, Tricia, threatened to write and remind him. But Beth had been warned by her old friend Ruth, Lady Crawford, whose husband, the Earl of Crawford and Balcarres, had also received an award that year, that the wheels sometimes moved slowly in royal circles. The invitation came eventually but by then Beth had already arranged to go with a group from the nursery, and she was not able to have a private visit after all. As a great admirer of the Prince and his views on organic gardening, she was also disappointed not to be able to go to Highgrove for a gathering of RHS Victoria Medal of Honour holders.

Beth's daily life was as busy as ever, with seemingly no let-up in people beating a path to her door and wanting a piece of her in one way or another. In one particularly busy Chelsea Flower Show week in May 2001, her diary was filled as follows: 'Tuesday 23 May, Helen Dillon with a film crew from RTE Irish TV; Wednesday 24 May, Thelma Barlow, "Mavis" from *Coronation Street*, with a film crew from Carlton TV, then entertained Sir Roy Strong and his wife, Julia Trevelyan Oman, together with Hugh Johnson and his wife, Judy, 5.30 p.m., note to watch TV programme *Inside Out*; Thursday 25 May, give an introductory talk to a group from Colman's of Norwich, dash to hairdressers at 11.45 a.m, another introductory talk in the afternoon to a group from Destination Europe; Friday 26 May, welcome talk to the Canterbury chapter of the New Zealand Horticultural Society, photo call for sitting with Tessa Traeger for the National Portrait Gallery, afternoon visit by Germaine Greer and her sister; Saturday 27 May, introductory talk to a group from the charity Plant Life.'

She had also been invited by Matthew Wilson, then curator of the RHS's Essex garden, Hyde Hall, to advise on their planting plan for a dry garden. Rather than see it as competition – Hyde Hall lies just thirty miles south-west of Elmstead Market – Beth was delighted, and happily agreed to open the garden in June on what ironically turned out to be one of the wettest days that summer.

This was also the year that a new gardener was taken on for an initial six-month term. Åsa Gregers-Warg had trained for a Masters in Gardening at Säbyholm in Sweden, during which time she also had done short stints at the Bergius Botanic Garden in Stockholm. After finishing her studies, she worked at a large nursery nearby before deciding she wanted to spend time working in an English garden. Åsa got to know Beth's garden and planting style through books and magazine articles and, as many had before her, she wrote to Beth asking if she could come and work for her for six months. Her quiet, dedicated style of working and passion for Beth's style of planting led to Beth offering her a full-time position. She never left, gradually taking on more of the responsibility of day-to-day gardening from Beth as she grew older. While everything was done under Beth's direction, increasingly she relied on Åsa, who Beth now called Senior Gardener, and her team in the

Beth with her old friend, confidant and writing companion Christopher Lloyd, with Christo's dachshund Dahlia sleeping soundly behind them.

garden, while David Ward continued to manage the nursery and Gerard Page the estate maintenance.

In April 2002, Beth arranged to take the team on a day trip to Dixter. For many, it was their first visit and a chance to see 'behind the scenes' of the garden they had all heard so much about. Christo's health was beginning to deteriorate but he was still up to teasing Beth even in front of her own staff. Knowing her distaste for bright colours, he took great delight in greeting the group in the Barn Garden standing in front of a bush of amber-coloured, bright orange spiraea which he had underplanted with pink hyacinths. With a wicked look in his eye, Christo expected Beth to be shocked by this combination, Fergus Garrett later remembered. Beth greeted and hugged him, then Christo pointed down to them and said, 'What do you think of this?' To the amusement of Garrett and Beth's assembled staff, she ignored him, walked on, pointed to a bergenia and calmly said, 'What a marvellous plant.' It was a rare occasion, Fergus thought, when Christo was left speechless.

While the pair continued to correspond and speak on the telephone, their age and health meant there could not be so many visits. Four years later, on 27 January 2006, her great friend died. Although Christo had had health problems, it was sudden and a shock.

It continued to be a difficult year. In late June, Beth fell and broke her knee. Later she had a shoulder joint replacement. While recuperating with her daughter Mary in Suffolk, she rang her twin brother, Seley, as she now did every day, to find him distressed. He was living in a care home in Colchester but was admitted to hospital later that day with a suspected heart attack. Beth and Mary drove to be with him and arrived in time for Beth to see him before he died that evening.

Although their lives and personalities had been quite different, Seley was always cheering Beth on from the wings, keeping her up to date with cuttings and clippings on her achievements and those of her friends. He had lived with their parents in Wivenhoe until their deaths, when he moved into a small house locally, from where he continued to be involved in the arts and reviewing for local newspapers. In contrast, Beth was, by her own admittance, never one to relax easily. In an interview she gave to the *Telegraph* magazine in 2007, she showed that not much changed since the fateful story in *The Sunday Times* magazine fifteen years previously. She still slept with the curtains open 'so I wake as soon as it gets light. That's usually about 5 a.m. in the summer . . . I don't get up immediately I wake. I cope with all the thoughts that come rushing into my head and plan my day.'

In June 2007, Penelope Hobhouse wrote to Beth asking whether she would consider being the subject of an exhibition at the Garden Museum in Lambeth, of which Hobhouse was a trustee. At that time, the museum was about to undergo a major transformation. The new director, Christopher Woodward, asked various well-known gardeners and garden writers which living gardener they would most like to see featured in an exhibition when the museum reopened in late 2008. Beth, Hobhouse wrote to tell her, had been the unanimous choice. It took a while for Beth to agree. Mary Guyatt, then the museum's curator, remembered visiting Beth and not being sure whether she had agreed or not – until she was introduced to a

RIGHT Beth with Dan Pearson and Christine Walkden, both founder patrons of the Beth Chatto Education Trust.
OPPOSITE Beth celebrating her ninety-fourth birthday on 27 June 2017 with her staff and granddaughter Julia Boulton.

member of staff as one of the people who was going to do an exhibition about Beth the following year.

More accolades followed. On Monday 18 August 2008, as part of the celebrations to mark Beth's eighty-fifth birthday that summer, BBC Radio 4's *Woman's Hour* broadcast a programme entirely devoted to her. It was, and remains, the only time a gardener had been given this rare honour.

In autumn 2008, to celebrate Beth's retrospective exhibition, the Garden Museum produced a special edition of their journal devoted to short essays on her written by her horticultural friends. Twenty-one contributed their 'thoughts' on Beth together with an interview by garden writer Diana Ross. Among them were several of Beth's former students including Doug Hoerr, Yuko Tanabe and Peter Janke and long-term friends Alan Titchmarsh, Fergus Garrett and Germaine Greer. The private view party for the exhibition was packed with Beth's friends and admirers from the horticultural world. 'Despite my fears over such a personal exposure,' she wrote later to her friend and former editor, Erica Hunningher, 'in

the event, the party was one of the happiest evenings of my life, so much warmth and sharing.'

But it was also a time of concern for Beth, since takings were down at the gardens and nursery and, for the first time, at one point in the year she was forced to borrow from the bank and put in some of her personal money to pay staff wages and other essential costs. While it was impressive that she had not had to do so before, it was an indication that the business was not keeping up with the adverse economic conditions. It had also been a particularly dull summer weather-wise. Every year Beth wrote a Christmas letter to all the staff, previously always with news of a small bonus. This year she was not able to do that, but she tried to remain optimistic. She finished by quoting a Chinese proverb: '"If a man has two coins, with one he buys a loaf, with the other a narcissus, to feed the body and the soul." Hopefully, there will be enough people this coming year who come to us for the narcissus.'

Visitor numbers and nursery sales did pick up and after the fiftieth anniversary year for the gardens, staff once more received a bonus, albeit a small one. There were also celebrations about the success of the gravel garden Beth had been asked to create at Marlborough House as part of Prince Charles's 'Garden Party to Make a Difference'. The ten-day festival was held across three royal London residences, Marlborough House, Clarence House and Lancaster House. Organised by the Prince's Charities Foundation to increase awareness of environmental issues and sustainability, it encompassed food, fashion, music and gardening, the last sponsored by the major British water companies. Beth was the natural choice to design a garden to demonstrate a planting scheme that would never need watering. She was given just six weeks to put plants together for the 32 × 32 foot space. With no hesitation, Beth produced a two-page handwritten 'shopping list' of plants that David Ward and the team needed to pot up for London, all plants that had survived at Elmstead without irrigation.

Also in 2008, a Japanese film crew lead by Hiroshi Fukumoto visited the garden four times to make a documentary about Beth and the garden to be shown on NHK, the national Japanese broadcasting network. Those who have seen the film (it has never been shown to the public outside Japan) found it had a rather otherworldly feel, with a small ghostly child in the garden and the occasional sheep. But that appealed to Beth, who felt that they were trying to discover the philosophy behind the garden – always a challenge for a journalist or filmmaker. Writing at a time of world recession, Beth reflected on the universal problem of making enough money to keep the garden and nursery going without changing too much her principles of caring, both for plants and people, 'many of whom find spiritual comfort and support in the garden'.

She went on to quote from a letter she had recently received from Isbert Preussler, her old friend Helen von Stein-Zeppelin's gardener, who had worked for many years in East Germany with the late Karl Foerster, 'a truly great plantsman', whose nursery celebrated its centenary in 2010. Preussler was saddened that nurseries were changing completely. 'Plants', he wrote with sadness, 'are not sensed as living creatures. The relationship between man and plant (nature) has changed, from peaceful co-operation to unilateral exploitation.' While Beth agreed, she felt differently about the

Beth remained active in the garden into her eighties and even nineties, always contributing suggestions for work needed.

future of the gardens. 'A garden is ephemeral,' she wrote. 'It is not a painting hanging on the wall, it is ever-changing. Who knows how it will develop after I am gone? I know my dear staff will do their best to carry on in the ways I would wish.'

The 2008 Beth Chatto exhibition at the Garden Museum was such a success that it persuaded Christopher Woodward that the museum ought to be collecting archives of great horticultural figures such as Beth. However, an archive would require specialist housing at huge expense. Woodward knew that having Beth's support was vital to the task of fundraising for the whole project. After some delicate negotiations, Beth was one of the first to agree to donate her archive, along with her friends John Brookes and Penelope Hobhouse. It was at this point that I became involved, helping Beth prepare her archive for its arrival at the museum when it reopened after its major extension in 2017. While she was no longer able to visit the museum, she was delighted to be able to support it in this way, and relieved that the responsibility for her archive's storage was someone else's, and it was no longer at the mercy of damp and mice while stored at the gardens.

THE LAST FEW YEARS of Beth's life were spent in the home she and Andrew had built in the centre of what was to become her world-famous garden. Hans Pluygers died in September 2016, aged ninety-four, having moved into Colchester on his retirement twenty years before. He and Beth stayed in touch but rarely saw each other, hearing news via a mutual friend. Beth did not attend his funeral.

Increasingly bed-bound, she would view the garden and its visitors through her bedroom window, occasionally receiving one of the endless stream of friends from across the world. In 2014, she wrote to a friend about the restructuring of the business. She was delighted and relieved that her eldest granddaughter, Julia Boulton, had come to take over the running of the nursery and gardens. 'Changes have to be made,' she explained, 'but Julia has the strength to do this.'

One of Julia's first tasks was the establishment of the Beth Chatto Education Trust, a charity to promote Beth's values and beliefs that education at all ages is key to an understanding of the world's ecological balance. The gardens had been running courses since 2006, including for RHS qualifications, but space was tight.

Although increasingly frail, Beth still wanted to be involved in all major decisions and, in 2016, she approved the building of a new teaching building. In summer 2017, she officially opened the Willow Room, tucked away by the damp plant propagation area, now giving access to a weatherproof classroom and a wildlife pond for visiting schoolchildren. Soon after, a visiting film crew from BBC's *Gardeners' World* was thrilled when she was well enough to be interviewed in the garden by presenter Carol Klein, who recalled her own vivid memories of visiting Beth's stands at Chelsea in the 1970s and 1980s.

Beth Chatto died at her home, White Barn House, on Sunday 13 May 2018. She had been in the gardens just the day before, taken out in her wheelchair, and as always noting what was coming up and what needed doing. Her funeral took place at the parish church of St Anne and St Laurence, where she had been married sixty-five years earlier. The church and Beth's seagrass coffin were decorated with flowers and foliage from the gardens she had created. She was buried with Andrew and her grave was strewn with more flowers, including many of the beautiful bearded irises bred by her great friend and mentor Cedric Morris, a perfect tribute to one of Britain's greatest and most influential gardeners.

17
Legend and legacy

On 13 July 2018, exactly two months after Beth's death, a celebration of her exceptional life was held at the Garden Museum in Lambeth. Britain was experiencing one of the longest heatwaves for years and temperatures soared to over 30 degrees, making travel around the city sticky and uncomfortable. The roads were grid-locked for another reason. President Trump (whose eponymous tower Beth had seen in 1983) was making his first official visit to London, prompting nearly a quarter of a million peaceful but passionate protesters to take to the streets.

But in the calm surroundings of the museum, a converted church nestling against Lambeth Palace, the Archbishop of Canterbury's London residence, a roll call of over a hundred leading horticulturists, garden designers and journalists gathered with Beth's family, friends and staff from the gardens to look back over her long and influential life. They included many of her former students who had gone on to have successful careers. On display were items from Beth's professional archive now in its new home at the museum – reminders of her staggering contribution to the way we think about plants and gardening.

That evening luminaries including Dan Pearson, Mary Keen, Matthew Wilson, Joy Larkcom and Fergus Garrett, lined up to pay tribute to Beth's revolutionary ideas. A number of themes emerged, as they had also done in the obituaries that appeared after her death in all the quality UK newspapers and gardening press. As Penelope Hobhouse – who, nearly thirty years before, had been with Beth, Christo and John Brookes at the symposium in Australia – wrote in her obituary in *The Guardian*: '[Beth's] approach was a revelation and immediately established her significance as a guide to better and more environmentally friendly gardening techniques.'

Britain's best-selling quality newspaper, the *Daily Telegraph*, published not only an obituary but also a double-page spread with tributes from some of those who had known her, among them garden photographer Jerry Harpur and TV presenter Carol Klein, as well as some of the world's leading garden designers including Piet Oudolf and Tom Stuart-Smith. In different ways, they all said that Beth's influence had changed the way they put plants together. They felt she inspired gardeners to forget the obsession with flowers and strong colour, encouraging them instead to use of the natural beauty of foliage and species plants in native settings, always considering their provenance, believing they would thrive given the right conditions in which to grow.

For many, Beth's impact started after her displays at the Chelsea Flower Show in the late 1970s and 1980s. Memorably causing a stir by showing species plants that some judges thought were 'weeds', she also broke the mould in how these plants were displayed by creating realistic garden groupings for different soil conditions. But her significance went deeper than that. The stands and even the show gardens one sees at Chelsea today all, to varying degrees, have been influenced by the way Beth ignored convention and brought her creative planting (inspired by ecology but also by her skill as a flower arranger and her fascination with the Japanese 'triangle of life' into the Great Marquee. Another repeated theme was that Beth changed the way gardeners look at plants, encouraging them to study their form, foliage and overall appearance throughout the year. Garden designer Sarah Price went so far as to say, 'She encouraged us to think like a plant.'

Beth applied her detailed knowledge to the plant descriptions in her *Handbook*, which was updated and reissued in 2015. Her entry for *Aconitum carmichaelii* 'Arendsii' not only tells the reader the plant's origins – Kamchatka and Amur Oblast – but also encourages the gardener to look closely at the plant. 'If you lift the top petal,' it says, 'you will find it is sheltering the stigma, while the rest of the petals form the familiar innocent face of the buttercup surrounding the central boss.'

As Beth frequently emphasised, none of this would have been possible without the input of Andrew Chatto. He inspired Beth's interest in the provenance of each plant and his prodigious research was the foundation on which she built her own encyclopaedic knowledge. But it was Beth's visual and creative talents, and her ability as a communicator, teacher and writer, that enabled her to pass on this knowledge. Tribute after tribute focused on not only Beth's skill at studying plants but also her talent for enthusing others to do the same. She was delighted that this would be carried on by the Beth Chatto Education Trust. Promoting horticultural

Discovered in the gardens: *Pulmonaria rubra* 'David Ward'.

education and an understanding of plant diversity in future generations lies at the heart of the gardens' mission.

But it is not just at the gardens that Beth's influence continues to be felt. Piet Oudolf, himself famous around the world for his planting schemes, sees Beth's impact as being truly international, having been crucial to the development of the New Perennial movement. At the Garden Museum tribute, James Hitchmough, professor of horticultural ecology at Sheffield University and mastermind behind the acclaimed naturalistic planting at London's Queen Elizabeth Olympic Park, said that Beth's garden was perhaps the most original British horticultural creation of the twentieth century. It will, he added, continue to have a profound effect on designers across the world in the twenty-first century.

An early success: *Kniphofia* 'Little Maid'.

The nursery remains a major source for some of Britain's leading gardens for rare forms which are often not available elsewhere. Many of these plants were discovered in the garden. One of the first successes was *Kniphofia* 'Little Maid'. More kniphofia introductions followed but none had the commercial success of 'Little Maid'. Often these finds resulted from the keen eyes of the skilled team Beth had working for her. David Ward was responsible for many discoveries; and one of Beth's propagators, Debbie Allcock, spotted two distinctive forms of *Gaura lindheimeri*, 'Jo Adela' and 'Corrie's Gold'. Beth was always generous in her acknowledgement of her plant sources – she brought back many fine examples, particularly from visits to Holland and Germany. Some of the plants that she introduced from Ernst Pagels's nursery in Leer, Ostfriesland, include *Achillea*

'Credo', *Aubrieta* 'Silberrand', *Papaver orientale* 'Leuchtfeuer', *Phlomis tuberosa* 'Amazone' and *Symphytum ibericum* 'Blaueglocken'. She was particularly struck by a number of forms of miscanthus selected by Pagels, many of which she made available to British gardeners through her nursery.

Other plants came to the nursery through her connections with Countess Helen von Stein-Zeppelin and Isbert Preussler, the Countess's head gardener. She was particularly fond of *Polygonatum* × *hybridum* 'Betberg', collected by Preussler and named after a site in the Black Forest.

In addition to her plant handbook, Beth's other books remain a vital source on ecological planting. Monty Don, gardening author and presenter of the BBC's *Gardeners' World* programme, in his introduction to the recent reprints of Beth's early books, *The Dry Garden*, *The Damp Garden* and *Beth Chatto's Notebook*, explains why he thinks this is so: 'When Beth Chatto wrote about plants . . . you knew that it was always from personal experience. This deeply practical, resourceful person was sharing her knowledge and advice rather than dispensing it from on high. There was always a modesty in this despite the rigour and remorseless attention to detail.'

It was this honest devotion to her plants that brought Beth so many fans across the world. She was sometimes puzzled by this adoration and it took a good friend to remind her why she had become such a legend. In 2011, Fergus Garrett, head gardener at Great Dixter, a garden that could not be more different from Beth's but which gave her great pleasure over the years through her friendship with Christopher Lloyd, brought a group of American visitors to Elmstead Market. Afterwards, he wrote her a thank you note explaining what it had meant to them all: 'Meeting you is a life-changing experience . . . and for those of us that know you well, you continue to be an inspirational light.'

Timeline

1923 Born in Good Easter, Essex
1943 Marries Andrew Chatto
1960 Family moves to White Barn Farm, Elmstead Market
1967 Opens Unusual Plants Nursery
1977 1st Gold Medal at the Chelsea Flower Show
1978 *The Dry Garden* published
1982 *The Damp Garden* published
1985 *Plant Portraits* published
1987 10th (and final) Chelsea Gold Medal
1987 RHS Victoria Medal of Honour
1987 RHS Lawrence Memorial Medal
1988 *Beth Chatto's Garden Notebook* published
1988 Honorary Doctorate, University of Essex
1989 *The Green Tapestry* published
1992 Planting begins in January on the Gravel Garden
1998 *Dear Friend and Gardener* (jointly with Christopher Lloyd) published
1998 Lifetime Achievement Award from the Garden Writers' Guild
1999 Andrew Chatto dies
2000 *Beth Chatto's Gravel Garden* published
2002 *Beth Chatto's Woodland Garden* published
2002 Awarded OBE
2004 *The Damp Garden* (new hardback edition with full colour photography)
2008 *Beth Chatto's Shade Garden* (previously published as *Beth Chatto's Woodland Garden*)
2009 Honorary Doctorate, Anglia Ruskin University
2011 Rotary Club Paul Harris Fellowship
2014 Society of Garden Designers John Brookes Lifetime Achievement Award
2018 Beth Chatto dies

Biographies

Brief notes on some people mentioned by Beth in her diaries and notebooks.

LeNeve 'Ollie' Adams (1931–2010) was a real estate agent in Raleigh, NC, who acquired her nickname from Olive Oyl in the Popeye cartoon when at college studying economics. She was a noted and creative gardener with an extensive knowledge of both rare exotics and native American plants. She travelled widely and led garden tours to the UK.

Paul Aden (1925–2010) was a schoolteacher with an early passion for daylilies (*Hemerocallis*). He later became a leading member of the American Hosta Society. After his death, the American Hosta Society reclassified the origins of many of his introductions as 'unknown'.

Mary Ashton is married to **Professor Peter Ashton**, who was director of the Arnold Arboretum and the Arnold Professor of Botany at Harvard University between 1978 and 1987. Born in England, Peter Ashton remained at Harvard as the Charles Bullard Professor of Forestry until his retirement in 2005.

Pierre Bennerup is CEO of the wholesale plant nursery Sunny Border Nurseries. Located in Kensington, CT, the nursery was founded by his father, Robert, in 1929. It specialises in perennial plants.

Marion Blackwell is an Australian ecologist and landscape architect involved in the creation of many of Australia's National Parks.

Alan Herbert Vauser Bloom (1906–2005) started a wholesale nursery business in the 1930s. After the war, it moved to his new home at Bressingham in Norfolk where it became one of the largest in the UK. In addition to introducing over a hundred perennial plants, he championed the idea of island beds, which became hugely popular in the 1970s and 1980s. **Adrian Bloom**, Alan Bloom's son, continued his father's work at Blooms of Bressingham. His garden at Bressingham, Foggy Bottom, became famous for his collection of conifers, which he promoted for all-year-round garden interest.

John Brookes MBE (1933–2018) was an influential and multi-award-winning British garden designer who pioneered modernist designs. His book *Room Outside* (1969) confirmed his position as a pre-eminent designer of 'outdoor rooms'. In 1980, he established a training school at Denmans, West Sussex.

Kathleen 'Kathie' Buchanan (1913–2010), of Keswick, VA, was an avid gardener and enthusiastic member of numerous garden clubs in Pennsylvania and Virginia as well as the Herb Society of America.

C. Colston 'Cole' Burrell is a garden designer, lecturer and author who lectures internationally on plants and ecology. His garden, Bird Hill in Charlottesville, VA, is known for its native planting, and has featured in many publications.

Myra, Lady Butter is a granddaughter of Grand Duke Michael of Russia and cousin of the Duke of Edinburgh. Her father, Major-General Sir Harold Wernher, inherited Luton Hoo in Bedfordshire.

Frank Cabot (1925–2011), chair of the New York Botanical Garden in the 1970s, started the not-for-profit Garden Conservancy. His 63-acre garden, Stonecrop, at Cold Spring, NY, opened to the public in 1992.

Edward 'Ted' Childs (1905–1996) was a distinguished forester and environmentalist, often called 'Connecticut's Teddy Roosevelt'. The family home was a 6,000-acre forest, Great Mountain Forest, at Norfolk, CT. After Childs's death, to protect the forest's future, his widow and family gave it to a non-profit trust.

John Codrington, Lt-Col., RHS Veitch Memorial Medal in 1989 (1898–1991), an English plantsman and artist, lived and gardened in Rutland. Self-taught, he designed gardens across the world from Emmanuel College, Cambridge, to Timbuktu. Among publications, he contributed to *Flora of Essex* in 1974. After their divorce in 1942, his wife, Primrose (née Harley), later married Lanning Roper (cf.).

Gordon Collier developed the renowned garden of Titoki Point in the centre of the North Island of New Zealand from the 1960s. Just 3 acres, it was one of the most visited gardens in the country but is no longer open to the public.

Ruth, Lady Crawford lives at Balcarres House, half a mile north of Colinsburgh near Fife. It has been in the Crawford family since the late sixteenth century. The gardens open for the Scottish Open Gardens Scheme.

Anthony 'Tony' du Gard Pasley (1929–2009) was a garden designer and landscape architect. For many years, he worked for Sylvia Crowe's London practice where John Brookes (cf.) was also working. He went on to lecture and teach widely, later becoming a principal judge of Chelsea Flower Show gardens.

Thomas 'Tom' H. Everett (1903–1986) worked at the New York Botanical Garden from 1932, retiring as director in 1968. Born in Woolton, Merseyside, he studied at the Royal Botanic Gardens, Kew, before moving to the US in the early 1930s. His magnum opus was the ten-volume *New York Botanical Garden Illustrated Encyclopedia of Horticulture* (1982).

Harry Lincoln Foster (1906–1989) wrote what is probably the most popular American book on rock gardening, *Rock Gardening: A Guide to Growing Alpines and Other Wildflowers in the American Garden* (1968). It was illustrated by his wife, Timmy, and described 1,900 plants in 400 genera.

Colin Hamilton and **Kulgin Duval** (1929–2016), rare book dealers, were introduced to Christopher Lloyd by Alan Roger (cf.), after which he regularly visited their home, Frenich, on the edge of Loch Tummel in Perthshire. Their extensive collection and archive have been donated to the National Library of Scotland and the National Galleries of Scotland.

Pamela Harper is an English gardening writer, photographer and lecturer who moved to America with her husband in the 1960s and is now based in Seaford, VA. She is best known for her books *Designing with Perennials* (1991) and *Color Echoes: Harmonizing Color in the Garden* (1994).

William 'Bill' Hean was head gardener and founding principal of the School of Heritage Gardening at Threave. The estate and school in Dumfries and Galloway is maintained by the National Trust for Scotland.

Douglas Henderson (1927–2007) was Regius Keeper of Britain's second oldest botanic garden (after Oxford), the Royal Botanic Garden, Edinburgh, between 1970 and 1987. On his retirement, he became the National Trust for Scotland's administrator at Inverewe Garden, Wester Ross.

Penelope Hobhouse MBE, VMH is a writer, garden designer and horticultural historian. Known first for her planting, especially her use of colour, at her Somerset homes at Hadspen and then Tintinhull (owned by the National Trust), she later wrote extensively on horticultural history, garden and planting design.

Major William George Knox Finlay (1895–1970) and **Mary Knox Finlay** (1897–1987) were both awarded medals by the RHS for their contributions to horticulture in developing the gardens at Keillour Castle.

Gary Koller was Supervisor of Living Collections and then Managing Horticulturist at the Arnold Arboretum in Boston, MA, between 1976 and 2000.

David and Melanie Landale's garden at Dalswinton remains in the family and opens for Scotland's Gardens Scheme.

(Charles) Brian and Margaret Anne 'Maggie' Lascelles's one-acre garden at The Bank House, in Glenfarg in Perthshire, was considered one of the best private gardens in Scotland. They maintained it until their retirement in 2008.

Arthur Lett-Haines (1894–1978) was an artist and sculptor, and co-founder with Sir Cedric Morris of the East Anglian School of Painting and Drawing at Dedham and later Benton End, near Hadleigh.

Joan Littlewood (1914–2002) was an English theatre director who is best known for her 1963 production of *Oh, What a Lovely War!* This was developed at the Theatre Royal Stratford East. In the early 1980s, she gave up directing and became companion to Baron Philippe de Rothschild.

Adele (née Brown) Lovett (1899– 1986) was married to **Robert A. Lovett** (1895–1986), the US Secretary of Defense. She came from a wealthy banking family and Robert Lovett joined their firm, Brown Brothers Harriman, in 1926.

Fred McGourty (1936–2006) was the nursery owner at Hillside Gardens, Norfolk, CT, and author of *The Perennial Gardener* (1989).

Marshall Olbrich (1920–1991) was an American plantsman who was a leading figure in the California Horticultural Society. He specialised in Mediterranean plants at his three-acre garden, Western Hills Gardens, Occidental, California.

Ernst Pagels (1913–2007) was a German gardener and plant breeder who worked with Karl Foerster. After the war, he founded a nursery in Leer, Ostfriesland, and was a prolific and influential perennial plant breeder.

Larry G. Pardue was director of the New York Horticultural Society between 1975 and 1988. In 1985 he reintroduced the New York Flower Show, after a fifteen-year hiatus. In 1988, he became the executive director of the Marie Selby Botanical Gardens in Sarasota, FA, and then, in 1991, director of the Los Angeles County Arboretum and Botanic Garden.

Jane Pepper was president of the Pennsylvania Horticultural Society for nearly thirty years, retiring in 2010. She oversaw the development of the Philadelphia Flower Show into an international show.

J. C. (James Chester) Raulston (1940–1996), a highly regarded horticulturist and plant collector, was director of the North Carolina State University Arboretum when he was killed in a car crash. The arboretum was renamed the J.C. Raulston Arboretum in his memory.

Alan Roger (1909–1997) gardened at the family home, Dundonnell, Little Loch Broom in Wester Ross. He was a close friend of Christopher Lloyd. Roger promoted the art of bonsai in the United Kingdom and exhibited regularly at the Chelsea Flower Show. He was also a judge on the RHS's Floral Committee 'B' and author of the RHS's booklet on bonsai.

Lanning Roper (1912–1983) was an American landscape architect, garden designer and horticultural writer who lived in England. In addition to work at Winston Churchill's home, Chartwell, Roper was commissioned by Prince Charles to help with the garden at Highgrove, but Roper died before he was able to do so. His wife, Primrose, had previously been married to John Codrington (cf.).

Philippine de Rothschild (1933–2014) was the only daughter of Baron Philippe de Rothschild. Her first husband was theatre director and actor Jacques Noël Sereys.

John Saladino is an American designer of interiors and gardens.

Lucy Landon Scarlett spent many years working at Longwood, her travels to South Africa and Brazil leading to the creation of the Silver Garden and the Cascade Garden. She left in 1989 to become director of the Dallas Arboretum and Botanical Garden.

Martha Schwartz is an American landscape architect and installation artist with bases in New York, London and Shanghai. She believes that the British obsession with private gardens has held back the development of public spaces in the UK.

George Seddon (1927–2007) was an Australian academic specialising in landscape diversity. He published widely but is best known for his book *Sense of Place* (1972) about the Swan Coastal Plain in Western Australia.

Robert 'Bob' and **Rosemary Seeley** were old friends of Christopher Lloyd's. Bob and Christo were contemporaries at Wye College after the Second World War and he regularly visited them at their home in the Tyne Valley.

Natasha Spender (1919–2010) was a pianist and author. She married Stephen Spender in 1941. She was a passionate gardener, and her book *An English Garden in Provence*, a memoir of the garden she created, was published in 1999.

Jim Sutherland is a Scottish plant collector, travelling to China, Nepal, Chile, Siberia and Europe in search of seeds. He started Ardfearn Nursery at Bunchrew near Inverness in 1987.

James van Sweden (1935–2013) was an American landscape and garden designer who with his partner Wolfgang Oehme pioneered the New American Garden Style, inspired by prairie-style planting of perennials and grasses.

David Tattersfield is an expert botanist and regularly leads wild-flower tours across the world from China to Ecuador.

Peter Thoday lectured in Landscape Management at the University of Bath for many years before becoming the Horticultural Director of the Eden Project.

Peter Valder is a botanist and author. An early interest in the native flora of Australia led to an academic career as a plant pathologist and mycologist.

Charles Webster (1906–1998), was president then chairman emeritus of the New York Horticultural Society between 1957 and 1991. He helped to inspire the community garden movement in New York City.

David Wheeler is founder and editor of the literary horticultural journal *Hortus*. His garden, Bryan's Ground at Presteigne, opens regularly to the public.

Ernest Henry Wilson (1876–1930), known as 'Chinese' Wilson, was a renowned British plant collector who introduced some two thousand plant species (of which around sixty were named for him).

Thomas William John 'Tom' Wright (1928–2016) was a much-loved Senior Lecturer in Landscape Management at Wye College, Kent, from 1978 to 1990. He first studied at Wye from 1949 to 1952, overlapping with Christopher Lloyd for the two years Lloyd was an assistant lecturer.

Index

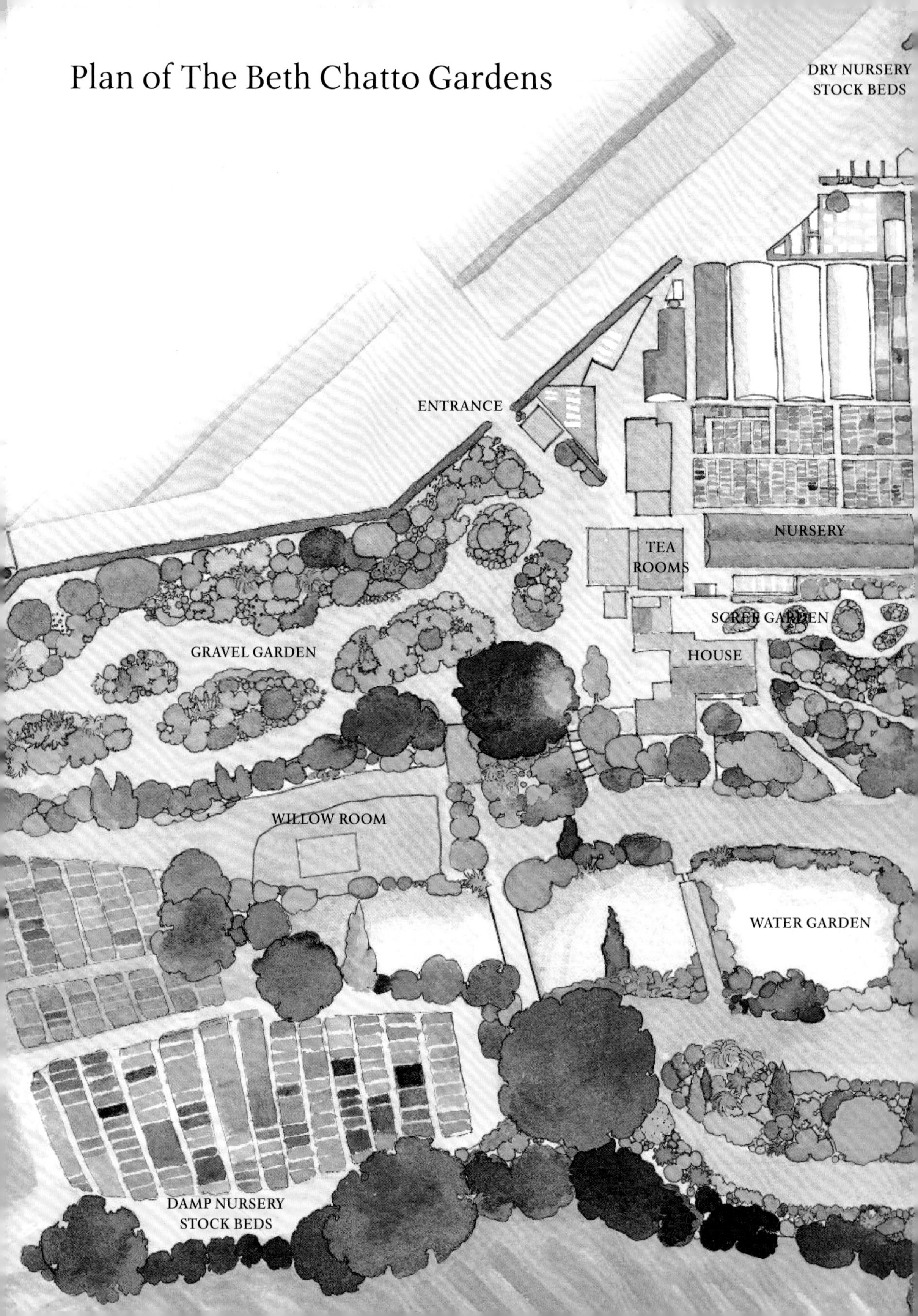
Plan of The Beth Chatto Gardens
DRY NURSERY
STOCK BEDS
ENTRANCE
NURSERY
TEA
ROOMS
SCREE GARDEN
GRAVEL GARDEN
HOUSE
WILLOW ROOM
WATER GARDEN
DAMP NURSERY
STOCK BEDS

WOODLAND GARDEN
RESERVOIR GARDEN
LONG SHADY WALK
E
ne
N
se
S
sw
W
nw

Acknowledgements

When I started work on this book, Beth, characteristically, gave me a list of friends she wanted me to contact. Of those I was able to speak to, I am especially grateful to the following, who so generously shared their memories of Beth with me and in many cases hospitality as well: Janet Allen, Stephen Anderton, Paul Baines, Adrian Bloom, Ronald Blythe, Helen Dillon, Professor Germaine Greer, Jerry Harpur, Professor James Hitchmough, Doug Hoerr, Peter Janke, Carol Klein, Piet Oudolf, Dan Pearson, Chrissy South, Yuko Tanabe, Bryan Thomas, Romke and Adriana van de Kaa, Christine Walkden, Susanne Weber, David Wheeler, Matthew Wilson, and Steven Wooster. Thanks also go in particular to Fergus Garrett for his guidance and wisdom, and his supporting staff at Great Dixter, Alan Titchmarsh for the loan of precious taped interviews and correspondence with Beth, and the late Erica Hunningher for her generosity in giving me her complete Beth Chatto archive. I am incredibly indebted to Christopher Woodward, director of the Garden Museum, who introduced me to Beth in the first place.

From our initial meeting, the enthusiasm for the book from Jo Christian and her publishing team at Pimpernel Press has been amazing. It has been a delight to work with them all. I am also extremely grateful to Tony Lord for his assistance with indexing. In addition, I could not have done without the practical help of Alice Lyzcia, Caroline Richardson and Hester Vickery Styles. As ever, my husband, Paddy Barwise, went beyond the call of marital duty, giving his editing skills during a 'busman's break' from writing his own book. And a special thank you to those members of the North Carolina Botanical Society who left no stone unturned in helping me piece together Beth's US canoeing trip.

Over the years, I have been extremely fortunate to have had the support of the staff, past and present, at the Beth Chatto Gardens. It is impossible to name them all but I am particularly grateful to Emily Allard, Tricia Brett, Åsa Gregers-Warg, Keith and Gerard Page, and David Ward. A special thank you to the staff in the Tea Room – my 'office' – who kept me going with endless cups of coffee and delicious 'Molly cake'!

Beth's family, in particular her daughter Mary and granddaughter Julia, have been enormously supportive throughout this project, always generous with encouragement, photographs and memories. I am grateful for their permission to quote from Beth's personal letters and diaries.

Finally, the greatest debt of thanks goes to Beth herself, for putting her trust in me to write her amazing story.

PHOTOGRAPH ACKNOWLEDGEMENTS

Unless otherwise noted below, images come from the Beth Chatto Estate and the photographic library of the Beth Chatto Gardens. Every effort has been made to identify individual photographers but please notify the publishers of any omissions.

Julia Boulton, p. 29; Helen Dillon, pp. 236, 260; the Garden Museum, p. 10; the Great Dixter Trust, pp. 125, 240; Douglas Hoerr, p. 206; Mary Marshall, pp, 38, 40, 41, 54, 55; Marion Nickig, p. 117; Piet Oudolf, p. 251, top and bottom; David Ward, p. 247; Rachel Warne, p. 2; Susanne Weber, p. 151; Ray Williams/Condé Nast Publications, p. 67.